The Besterman World Bibliographies

*The Besterman World Bibliographies*

# Agriculture

## A BIBLIOGRAPHY OF BIBLIOGRAPHIES

By Theodore Besterman

TOTOWA, N. J.

*Rowman and Littlefield*

*1971*

Published by
Rowman and Littlefield
A Division of
Littlefield, Adams & Co.
81 Adams Drive
Totowa, N. J. 07512

★



★

Typography by George Hornby
and Theodore Besterman

ISBN 0-87471-049-9

*Contents*

# Preface

I have explained in the Introduction to the successive editions of *A World bibliography of bibliographies* why I decided to arrange it alphabetically by specific subjects. Since that decision was taken, and after prolonged experience of the book in use, I have had no reason to regret it, nor among the many letters I have received from librarians has there been a single one complaining of the alphabetical form of the *World bibliography*.

The *World bibliography of bibliographies* covers all subjects and all languages, and is intended to serve reference and research purposes of the most specific and specialised kind. Yet contained in it are broad and detailed surveys which, if relevant entries throughout the volumes are added to them, can serve also the widest reference inquiries, and be useful to those who seek primary signposts to information in varied fields of inquiry.

Therefore I can only thank Rowman and Littlefield for having gathered together all the titles in some of the major fields found throughout the 6664 columns of the fourth edition (1965-1966) of *A World bibliography of bibliographies*.

# *Preface*

These fields are:

1. Bibliography
2. Printing
3. Periodical Publications
4. Academic Writings
5. Art and Architecture
6. Music and Drama
7. Education
8. Agriculture
9. Medicine
10. Law
11. English and American Literature
12. Technology
13. Physical Sciences
14. Biological Sciences
15. Family History
16. Commerce, Manufactures, Labour
17. History
18. Geography

Of course these categories by no means exhaust the 117,000 separately collated volumes set out in the *World bibliography*, and the above titles will be added to if librarians wish for it.

Th.B.

# *Notes on the Arrangement*

### *An Alternative to critical annotation*

Consider what it is we look for in a normal bibliography of a special subject. Reflection will show, I think, that we look, above all, for completeness, just as we do in a bibliography of bibliographies. We desire completeness even more than accuracy (painfully uncongenial though it is for me to make such a statement); for in most cases a bibliography is intended to give us particulars of publications to which we wish to refer; thus we can always judge for ourselves (waiving gross errors) whether the bibliographer has correctly described these publications. On the other hand, anything that is omitted is lost until rediscovered.

The question is, therefore, whether it is possible to give some indication of the degree of completeness of a bibliography without indulging in the annotation which is impossible in a work of the present scope and scale. It seemed to me that this could be achieved, to a considerable extent, by recording the approximate number of entries set out in it. This method is, of course, a rough-and-ready one, but experience shows that it is remarkably effective: and I hope that its novelty will not tell against it.

The recording of the number of works set out in a bibliography has another advantage in the case of serial publications: it displays in statistical form the development of the subject from year to year—often in a highly significant manner.

This procedure, then, is that which I have adopted, the number of items in each bibliography being shown in square brackets at the end of the entry. This, I may add, is by no means an easy or mechanical task, as can be judged from the fact that this process, on the average, just about doubles the time taken in entering each bibliography.

### *Supplementary information in footnotes*

I have said that this method of indicating the number of entries is intended to replace critical treatment; but it is not possible to exclude annotation altogether, for a certain minimum of added information is indispensable. Consequently many of my entries will be seen to have footnotes, in which the following types of information are recorded: a few words of explanation where the title is inaccurate, misleading, obscure, or insufficiently informative; a statement to that effect where a work is in progress, where intermediate volumes in a series have not been published, or where no more have been published; an attempt to clarify complicated series; a note that a book was privately printed or in a limited number of copies, where this does not exceed 500, or in some abnormal manner, as on one side of the leaf, on coloured paper, or in a reproduction of handwriting, or with erratic pagination; when I have come across copies containing manuscript or other added matter, I have recorded the fact; substantial corrections and additions to bibliographies are sometimes published in periodicals, and I have noted a good many of these—but without aiming at anything even remotely approaching completeness, the attainment of which would be impossible. Various minor types of information are also occasionally to be found in the footnotes.

Owing to the great increase in the number of bibliographies reproduced directly from typewritten copy, such publications are designated by an asterisk at the end of the entry; this device saves a good deal of space.

### *Place of publication*

The place of publication is not shown when it is London in the case of an english book and

Paris in the case of a french one. In the case of a series or sequence of entries, however, the absence of a place of publication means that it is the same as the place last shown in the series. The same applies to the names of editors and compilers. The place of publication is given as it appears on the titlepage, but prepositions are omitted even if violence is done to grammatical construction.

### *The Order of entries*

Under each heading the order of the entries is chronological by date of publication; in the case of works in successive volumes or editions the chronological order applies to the first volume or edition. In suitable long headings an additional chronological order by period covered has been created; see, for instance, France: History, or Drama: Great Britain.

### *Method of collating*

An effort has been made, so far as space allows, to give detailed and accurate information of the kind more usually found in small bibliographies. For instance, I have paid special attention to the collation of bibliographies in several (or even numerous) parts or volumes. It is, in fact, difficult to understand why it is usually considered necessary to give collations of works in a single volume, where difficulties seldom occur (from the point of view of systematic bibliography), but not of a work in several volumes, where confusion much more frequently arises. An occasional gap in the collations of such publications will be noticed. This is because, naturally enough, I have not been able in every case to see perfect sets; and I have thought it better to leave a very small number of such blanks rather than to hold up the bibliography indefinitely.

### *Serial publications*

Where successive issues of a serial publication are set out, the year or period shown is usually that covered by the relevant issue; in such cases no date of publication is given unless publication was abnormal or erratic in relation to the period covered.

### *Bibliographies in more than one edition*

Where a bibliography has gone into more than one edition I have tried (though I have not always been able) to record at least the first and latest editions. Intermediate editions have also been recorded wherever it seemed useful to do so, that is, for bibliographies first published before 1800, and for those of special interest or importance; but in general intermediate editions, though examined, have not been recorded.

# Notes

### *Transcription of titles*

Titles have been set out in the shortest possible form consistent with intelligibility and an adequate indication of the scope of the bibliography; omissions have of course been indicated. The author's name, generally speaking, is given as it appears on the titlepage, amplified and interpreted within square brackets where necessary.

### *Anonymous bibliographies*

Far too large a proportion of bibliographical work is published anonymously. This is due, in part, to the all too common practice of library committees and similar bodies of suppressing altogether or of hiding in prefaces the names of those who have compiled bibliographies and catalogues for them. I have spent a good deal of time in excavating such and other evidences of authorship, and the result may be seen in the large number of titles preceded by names enclosed within square brackets.

Th. B.

## Agriculture.

*Agriculture*

## 1. *Bibliographies*

E. LUCY OGDEN and EMMA B[EATRICE] HAWKS,

List of manuscript bibliographies and indexes in the U. S. department, including serial mimeographed lists of current literature. Department of agriculture: Library: Bibliographical contributions (no.11): Washington 1926. ff.iii.38. [202.]*

[VICTOR A. SCHAEFER *and others*], Aperçu des bibliographies courantes concernant l'agriculture et les sciences connexes. Institut international d'agriculture: Rome 1937. pp.[v].84. [225.]

— — Nouvelles sources bibliographiques agricoles. [By Sigmund von Frauendorfer]. 1943. pp.3–44. [100.]

[VIRGINIA MICHAELIS], Aids for library research in agriculture. University of Georgia: College of agriculture: [Athens 1955]. ff.[i].19. [100.]*

MARIE VOSÁHLOVÁ, Soupis nejnovějších zeměděslkých bibliografií. Universita: Knihovna: Publikace (no.3/55): Olomouc 1955. ff.[i].25. [50.]*

P. S. BOGOMOLOVA, Материалы к библиографии сельскохозяйственной литературы. Государственная публичная библиотека имени М. Е. Салтыкова-Щедрина. Ленинград 1955. pp.[i].84. [50.]

RUDOLF LAUCHE, Internationales handbuch der bibliographien des landbaues. Herausgegeben im auftrag des Land- und forstwirtschaftlichen forschungsrates. München 1957. pp.xvi.411. [3887.]

BIBLIOGRAPHIC aids for literature research. University of Minnesota: Institute of agriculture: Library: St Paul [1957]. pp.[ii].21. [100.]*

JEAN HASSENFORDER, La documentation en économie rurale. Quelques indications sommaires à l'usage des étudiants. Centre d'études économiques: 1958. ff.12. [30.]*

GHID de documentare în ştiinţele agricola. Academia Republicii Populare Romîne [&c.]: [Bucharest] 1960. pp.271. [2313.]*

## 2. *History*

### i. *General*

WILLIAM TRIMBLE, Introductory manual for the study and reading of agrarian history. [North Dakota agricultural college: Department of history and social science: Fargo 1917]. pp.47. [500.]

CATALOGUE of early works on botany, agriculture and horticulture. Public library: Cardiff 1918. pp.24. [60.]

CAROLINE HUBBARD BAILEY, The romantic and

historic background of agriculture and plant study. Public library: Bulletin (no.175): Riverside, Cal. 1921. pp.39. [200.]

MARY S. ASLIN, Catalogue of the printed books on agriculture published between 1471 and 1840. Rothamsted experimental station: Rothamsted [1926]. pp.332. [5000.]

— — Second edition. Library catalogue of printed books and pamphlets [&c.]. Harpenden 1940. pp.204. [4000.]

GUIDE to an exhibition of manuscripts and printed books illustrating the history of agriculture. British museum: 1927. pp.30. [91.]

G[EORGE] E[DWIN] FUSSELL, Chronological list of early agricultural works in the library of the Ministry of agriculture and fisheries. 1930. pp.43. [400.]

BIBLIOGRAPHY on the history of early agriculture, with special reference to the introduction of implements. Science library: Bibliographical series (no.52): [1932]. ff.[i].2. [28.]*

EVERETT E[UGENE] EDWARDS, Referencces on agricultural history as a field for research. Department of agriculture: Bibliographical contributions (no.32): Washington 1937. pp.v.41. [75.]*

*Agriculture*

KENNETH MAC LEISH and HELEN E. HENNEFRUND, Anthropology and agriculture. Selected references on agriculture in primitive cultures. Compiled . . . under the direction of Mary G. Lacy. Department of agriculture: Bureau of agricultural economics: Agricultural economics bibliography (no.89): Washington 1940. pp.xi.134. [270.]*

E[DOARDO] VOLTERRA, Saggio bibliografico di diritto agrario romano. Osservatorio italiano di diritto agrario: Testi legislativi e documenti: Firenze 1951. pp.141. [1450.]

ii. *Countries &c.*

*America*

EVERETT E[UGENE] EDWARDS, Agriculture of the american Indians. A classified list of annotated historical references. Edition 2. Department of agriculture: Bibliographical contributions (no.23): Washington 1933. pp.xii.106. [524.]*

EVERETT E. EDWARDS and WAYNE D. RASMUSSEN, A bibliography on the agriculture of the american Indians. Department of agriculture: Miscellaneous publication (no.447): Washington 1942. pp.v.107. [841.]

## *Agriculture*

### *China*

LIOU-HO and CLAUDIUS ROUX, Aperçu bibliographique sur les anciens traités chinois de botanique, d'agriculture, de sériculture et de fungiculture. Lyon 1927. pp.39. [75.]

### *France*

[NICOLAS] GEORGES [MARIE] BOURGIN, Inventaire de la sous-série $F^{10}$ [Archives nationales]. Société de l'histoire de la révolution française: 1918. pp.xxv.357. [large number.]

### *Great Britain*

DONALD MCDONALD, Agricultural writers, from sir Walter of Henley to Arthur Young, 1200–1800. 1908. pp.[v].228. [500.]

H. L. BEALES and J. D. CHAMBERS, A select bibliography of rural economic history. Association of tutors in adult education: Book lists (new ser., no.1): Edinburgh [printed] 1933. pp.[iv].14. [150.]

EVERETT E[UGENE] EDWARDS, Selected references on the history of english agriculture. Department of agriculture: Bibliographical contributions (no.24): Washington 1935. pp.v.42. [55.]*

— — (Edition 2). 1939. pp.v.105. [112.]*

G[EORGE] E[DWIN] FUSSELL, The old english

farming books from Fitzherbert to Tull, 1523 to 1730. 1947. pp.[vii].142. [350.]

— More old english farming books from Tull to the Board of agriculture, 1731 to 1793. 1950. pp.ix.187. [350.]

[A. ANDERSON], Catalogue of the Walter Frank Perkins agricultural library. Southampton university: Library: Southampton 1961. pp.xii.291. [2009.]

*Netherlands*

B[ARTHOLOMEUS] W[ILLEM] A[NNE] E[LISA] SLOET TOT OLDHUIS and W. J. D. VAN ITERSON, Proeve eener opgave van bouwstoffen voor eene geschiedenis en statistiek van den nederlandschen landbouw. 27e Nederlandsch landhuishoudkundgi congres: 's-Gravenhage 1874. pp.146.

*Poland*

MATTEW M. FRYDE, Selected works on polish agrarian history and agriculture. A bibliographical survey. Mid-european studies center: New York 1952. pp.[vii].87. [486.]*

*Sweden*

P[ER] M[AGNUS] HEBBE, Den svenka lantbruks-litteraturen.

i. Från äldsta tid t.o.m. år 1800. Uppsala 1939. pp.xvi.333. [2194.]

ii. Från 1801 till 1850. . . . Efter författarens efterlämnade anteckningar utgiven av C. Björkbom. Stockholm 1945. pp.xvi.372. [2970.]

*United States*

LIST of references on soldiers' and sailors' homestead legislation after the civil war. Library of Congress: [Washington] 1917. ff.4. [25.]*

EVERETT E[UGENE] EDWARDS, A bibliography of the history of agriculture in the United States. Department of agriculture: Miscellaneous publication (no.84): Washington 1930. pp.iv.307. [4273.]

EVERETT E[UGENE] EDWARDS, Selected references on the history of agriculture in the United States. Department of agriculture: Bibliographical contributions (no.26): Washington 1935. pp.v.28. [34.]*

EVERETT E[UGENE] EDWARDS, References on american colonial agriculture. Department of agriculture: Bibliographical contributions (no. 33): Washington 1938. pp.v.101. [257.]*

### 3. *Manuscripts*

HAANDSKRIFTER i den Kongelige veterinaer- og landbohøjskoles bibliotek. København 1907. pp.26. [190.]

### 4. *Periodicals*

#### *General*

VERZEICHNIS der periodischen schriften in der bibliothek der Ökonomischen gesellschaft im königr. Sachsen. Dresden 1892. pp.31. [300.]

PERIODICALS and society publications currently received at the department library. Department of agriculture: Library bulletin (no.2): [Washington] 1894. pp.8. [450.]

— [another edition]. List of serials currently received [&c.]. Department of agriculture: Department circular (no.187): 1922. pp.v.358. [5586.]

— [another edition]. By Elizabeth G. Hopper. . . . Department of agriculture: Miscellaneous publication (no.245): 1936. pp.iv.337. [3871.]

— [another edition]. Compiled by Katharine Jacobs. Department of agriculture: Bibliographical bulletin (no.12): 1950. pp.[ii].349. [14,329.]*

— [another edition]. Compiled by Elizabeth Gould Davis. Department of agriculture: Miscellaneous publications (no.765): 1958. pp.[ii].614. [22,500.]*

*other intermediate editions were published.*

BIBLIOGRAPHIE der gesamten landwirtschaftlichen zeitschriftenliteratur. Institut international de bibliographie [Bruxelles]: Contribution (no. 59): Vienna [1907] pp.112. [3200.]

LISTE des revues et journaux régulièrement reçus par l'Institut jusqu'au 1er avril 1912. Institut international d'agriculture: Rome 1912. pp.84. [2520.]

— [another edition]. Liste des périodiques reçus couramment par la bibliothèque de l'Institut international d'agriculture. Édition provisoire. 1935. ff.[i].v.133. [2686.]*

CATALOGUE of periodical and serial publications filed in the library of the Board of agriculture. Board [*afterwards:* Ministry] of agriculture and fisheries: Miscellaneous publications (no.14): 1913. pp.80. [1250.]

[WANT list of periodicals]. Experimental station: Rothamsted [1914]. pp.4. [100.]

[—]. [1915]. pp.4. [100.]

СПРАВОЧНИКЪ по сельскохозяйственной періодической печати на 1915 г. Петроградъ 1915. pp.[vi].iv.xxx.216. [310.]

## *Agriculture*

CATALOGUE of journals and periodicals. Rothamstead experimental station: [Harpenden] 1921. pp.70. [1000.]

— [another edition]. 1932. ff.[i].66. [1000.]*

LIST of current agricultural periodicals and allied serials on file in the library of the Bureau of science. [Manila] 1928. ff.[iii].4.16. [300.]*

IVA OLIVER and ZELLA RANK, List of serials in the Colorado state agricultural college library. Fort Collins 1930. pp.ii.68. [1250.]*

ZEITSCHRIFTEN-verzeichnis der bibliothek und der anderen institute der Landwirtschaftlichen hochschule Hohenheim. Hohenheim 1932. ff.80. [800.]*

LIST of periodicals and reports, &c. (including herd, flock, and stud books) in the library of the school of agriculture, Cambridge. Imperial bureau of plant genetics [*afterwards:* Commonwealth bureau of plant breeding and genetics]: 1932. ff.86.[i].3. [1000.]*

— Supplement. [1932]. ff.[3]. [21.]*

— Supplement no.2. [1933]. ff.[5]. [42.]*

— Supplement no.3. [1933]. ff.6. [100.]*

PERIODICALS containing agricultural references, issued in countries other than the United States and Canada. Science library: Bibliographical series (no.110): [1933]. ff.[i].2. [38.]*

PERIODICALS and books relating to tropical agriculture and forestry. Science library: Bibliographical series (no.75): [1933]. ff.[i].6. [98.]*

RELACIÓN de las distintas publicaciones periódicas que se reciben en la hemeroteca del Servicio de publicaciones agrícolas y biblioteca. Dirección general de agricultura: [Madrid] 1934. pp.[48]. [630.]

MAGAZINES contributing most frequently to the index of Current literature in agricultural engineering. Department of agriculture: Bureau of agricultural engineering: Washington] 1934. pp. [ii].17. [175.]*

OLGA KUKUTSCH, Poppelsdorfer zeitschriftenkatalog. Verzeichnis der in der hauptbücherei und den institutsbibliotheken der Landwirtschaftlichen fakultät der universität Bonn vorhandenen zeitschriften und reihen. Bonn 1935. pp.vii.106. [1203.]

— — 2. auflage. Zeitschriftenkatalog der universitätsbibliothek, abt. landwirtschaft und der landwirtschaftlichen institute der universität Bonn. Von Olga Kukutsch Anders. Bonner beiträge zur bibliotheks- und bücherkunde. (vol.2): Bonn 1956. pp.94. [1090.]*

CATÁLOGO de publicaciones periódicas. Universidad nacional: Facultad de agronomía y veterineria: Biblioteca: Buenos Aires 1937. pp.28. [659.]

VERZEICHNIS der laufenden zeitschriften der landwirtschaftlichen und tierärztlichen bibliotheken in Berlin. Berlin 1936. pp.iii.29. [1103.]*

NAMES and addresses of some of the chief agricultural journals issued in various european countries. Science library: Bibliographical series (no.307): 1937. ff.5. [70.]*

[F. H. BOLT], Catalogue of the periodical literature filed in the library, Department of agriculture, New South Wales. Sydney 1938. pp.132. [3000.]

IDA KEELING CRESAP, Catalog of periodicals in the library, Florida agricultural experiment station, Gainesville, Florida. [Gainesville] 1938. ff.[ii].42. [615.]*

CAROLYN WHITLOCK, Abbreviations used in the Department of agriculture for titles of publications. Department of agriculture: Miscellaneous publication (no.337): Washington 1939. pp.278. [7500.]

*refers to periodical publications.*

RÉPERTOIRE international de la presse agricole. Fédération internationale de la presse agricole: Rome 1939. pp.419. [4091.]

[SIGMUND VON FRAUENDORFER *and others*], Catalogue des périodiques possédés par la bibliothèque. Institut international d'agriculture: Rome 1946. pp.iii–xix.568. [5572.]

*PUBLICAÇÕES periódicas estrangeiras inventariadas nas bibliotecas portuguesas. Etnologia, ciências naturais, agro-pecuária. Instituto de alta cultura: Centro documentação científica: Lisboa 1953. pp.739. [8000.]

J. BARTIER, Répertoire des périodiques agricoles existant dans les principales bibliothèques spécialisées de Belgique. Commission belge de bibliographie: Bibliographia belgica (no.12): Bruxelles 1955. pp.125. [1527.]

LØPENDE tidsskrifter. Liste over løpende tidsskrifter og noen viktige årsskrifter i Landbrukshøgskolens hovedbibliotek og instituttbibliotekene. [Aas 1956]. ff.iv.70. [1100.]*

## *Agriculture*

LARS FRYKHOLM, Förteckning över löpande utländska lantbruksserier i svenska bibliotek. Uppsala 1956. pp.[iv].87. [3000.]*

— — Supplement 1958. Av Hans Baude. pp.[iv].30. [1100.]*

TH. P. LOOSJES [*and others*], Wageningse tijdschriften lijst bevattende een opgave van tijdschriften en series aanwezig in bibliotheken op het gebied van de landbouw te Wageningen en omstreken. Landbouwhogeschool: Bibliothek: Wagening 1958–1959. pp.x.417. [12,500.]*

— — Supplement. 1959. pp.[iii].30. [900.]*

LIST of periodicals etc. held in the library. Fifth edition. University of Oxford: Department of agriculture: Library: Oxford 1959. ff.[i].26. [500.]*

## *Countries &c.*

### *America, latin*

LIST of agricultural periodicals, societies, experiment stations, and schools in latin America. Pan american union: Division of agricultural cooperation: Washington 1941. ff.[iii].58. [408.]*

## *Agriculture*

ELIZABETH G. HOPPER, *ed.* A preliminary list of Latin american periodicals and serials. Department of agriculture: Library: Library list (no.5): Washington 1943. pp.195. [2000.]*

## *Bolivia*

ARMANDO CARDOZO [GONZÁLES], Revistas agrícolas bolivianas. Cronología y classificación. Sociedad de ingenieros agrónomos de Bolivia: Boletín bibliográfico (no.1): La Paz 1962. pp. [iv].ff.69.[vi]. [1808.]*

## *Canada*

H. J. ARSENAULT, Canadian agricultural serials currently received. Department of agriculture: Library: Library list (no.4): Ottawa 1962. pp.[ii]26. [500.]*

## *China*

LESLIE T. C. KUO and PETER B. SCHROEDER, Communist chinese periodicals in the agricultural sciences. Department of agriculture: Library list (no.70): Washington 1960. pp.24. [115.]*

## *Europe*

RÉPERTOIRE des quotidiens et périodiques agricoles dans les pays membres de l'O.E.C.E. Agence européenne de productivité: 1960. pp.144. [1500.]

## *Agriculture*

### *France*

[F. HOUILLIER and CH. POUPEAU], Enquête sur la presse agricole, 1952–1953. Ministère de l'agriculture: Service d'études et de documentation: [1953]. pp.121. [1000.]

*limited to french periodicals.*

### *Germany*

CHRISTIAN JENSSEN, Zur geschichte der landwirtschaftlichen zeitschriften Deutschlands von der mitte des vorigen jahrhunderts bis auf die gegenwart. Berlin 1889. pp.118. [245.]

DIE FACHPRESSE der land-, forst- und ernährungswirtschaft. Hamburg [1960]. pp.204. [603.]

### *Great Britain*

F. A. BUTTRESS, Agricultural periodicals of the British isles, 1681–1900, and their location. University of Cambridge: School of agriculture: Cambridge 1950. pp.16. [500.]

### *Hungary*

MEZOGAZDASÁGI szäkfolyóirataink bibliografiája. Az agrártudományi egyetem központi könyvtaranak kiadvanyai (4th ser., no.3): Budapest.

i. 1796–1899. Összeállitotta Turányi Kornél. 1958. pp.116. [318.]
ii. 1900–1944. Összeállitotta Mados Lászlóné. 1961. pp.175. [506.]

*Japan*

[PETER B. SCHROEDER], Japanese serial publications in the National agricultural library. Department of agriculture: National agriculture library: Library list (no.72): Washington 1962. pp.iii.172. [1118.]*

*Russia*

TRANSLITERATED and translated lists of Russian periodicals in the library of the School of agriculture. Cambridge 1936. ff.[iii].11.11. [80.]*

ПЕРИОДИЧЕСКАЯ печать СССР, 1917–1949. Библиографический указатель. Журналы, труды и бюллетени по сельскому хозяйству. Москва 1955. pp.230. [1947.]

*Sweden*

LARS FRYKHOLM, Förteckning över löpande svenska lantbruksserier med uteslutning av frökataloger, skolredögorelser och. div. lokala organisationers årsberättelser. Uppsala 1957. ff.12. [300.]*

## *Agriculture*

HANS BAUDE, Classified list of swedish periodicals in the fields of agriculture, fishery, forestry, veterinary medicine and allied subjects. Uppsala 1959. ff.[ii].34. [300.]*

### *United States*

GEORGE BATTEN's directory of the agricultural press. New York 1895. pp.[iv].93.13. [500.]

— [another edition]. Batten's agricultural directory. A directory of the agricultural press of the United States and Canada. [1908]. pp.270. [500.]

GILBERT M. TUCKER, American agricultural periodicals. Albany, N.Y. 1909. pp.[ii].71–80. [150.]

*privately printed from plates originally intended for an encyclopedia.*

ESTHER M[ARIE] COLVIN, List of state official serial publications containing material on agricultural economics. Department of agriculture: Bureau of agricultural economics: Agricultural economics bibliography (no.38): Washington 1932. pp.[ii].222 [*sic*, 223]. [205.]*

STEPHEN CONRAD STUNTZ, List of the agricultural periodicals of the United States and Canada published during the century July 1810 to July 1910. . . . Edited by Emma B[eatrice] Hawks.

Department of agriculture: Miscellaneous publication (no.398): Washington 1941. pp.viii.190. [3753.]

PARTIAL list of United States farm papers received in the U. S. Department of agriculture. Department of agriculture: Library: Library list (no.18): Washington 1945. pp.6. [60.]*

LIST of "agricultural outlook" publications prepared by state extension service. Assembled by Farm management and general economics section, Division of agricultural economics. Washington [1953]. pp.6 [125.]

SELECTED list of american agricultural books in print and current agricultural periodicals. Department of agriculture: Library list (no.1): Washington 1960. pp.34. [1250.]

### 5. *General*

JULIUS BERNHARD VON ROHR, Compendieuse haushaltungs-bibliotheck, darinnen nicht allein die neuesten und besten autores, die so wohl von der hausshaltung überhaupt, als von ackerbau, viehzucht, jägerey, gärtnerey . . . u.s.w. geschrieben, recensiret. Leipzig 1716. pp.[xiv].472.[xxii]. [750.]

— — Dritte . . . auflage. 1755. pp.[xxvi].492. [xvi.] [1000.]

[BARON OTTO VON MÜNCHHAUSEN], Des hausvaters zweyten theils erstes stuck. Inhalt: des hausvaters botanische, physikalische und oekonomische bibliothek. Hannover 1765–1766. pp.[ii]. xxxvi.[xii].367+369–832.[ciii]. [4000.]

[—] — Zweite . . . auflage. 1774–1775. pp. xxxvi.[xii].400+401–976. [5000.]

*reissued in 1784 as a so-called 'dritte . . . auflage'.*

JOHANN TRAUGOTT MÜLLER, Einleitung in die oekonomische und physikalische bücherkunde und in die damit verbundenen wissenschaften bis auf die neuesten zeiten. Leipzig 1780–1784. pp. [xvi].560+[ii].viii.718+[iv].iv.743. [15,000.]

J[EREMIAS] D[AVID] REUSS, Repertorium commentationum a societatibus litterariis editarum. . . . Tom. VI. Oeconomia. Gottingae 1806. pp. xvi.476. [5000.]

*each part of this work is entered under its appropriate subject; a facsimile of the whole was published New York 1962.*

[COUNT] FILIPPO RE, Dizionario ragionato di libri d'agricoltura, veterinaria, e di altri rami d'economia campestre, ad uso degli amatori delle cose agrarie e della gioventù. Venezia 1808–1809. pp.324+350+330+315. [3000.]

[VICTOR DONATIEN DE MUSSET-PATHAY], Bibliographie agronomique ou dictionnaire raisonné des ouvrages sur l'économie rurale et domestique et sur l'art vétérinaire. 1810. pp.[iii].xxiv.459. [2078.]

КАТАЛОГЪ библіотеки Императорскаго вольнаго экономическаго общества. Санктпетербургъ 1846. pp.[iv].323. [3000.]

HUGO EMIL SCHOBER, Encyclopädie der landwirtschaftswissenschaft. Nebst einer uebersicht über die neuere deutsche landwirtschaftliche literatur. Dresden 1856. pp.[ii].x.171. [400.]

KARL ANDRÉ, Die landwirtschafts-literatur der letzten dreizehn jahre. Prag 1860. pp.98. [1350.]

LOUIS BOUCHARD [-HUZARD], Bibliographie. Ouvrages publiés jusqu'à ce jour sur les constructions rurales et sur la disposition des jardins. 1860. pp.[iv].825–876. [275.]

*50 copies printed.*

— — Deuxième édition. 1870. pp.[iv].947–1048. [500.]

*90 copies printed; this work also forms part of the author's* Traité des constructions rurales.

CATALOGO della biblioteca. Ministero di agricultura, industria e commercio: Roma 1860 &c.

JOSÉ CANALEJAS Y CASAS, Annuario de los progresos tecnológicos de la industria y de la agricultura. Madrid 1861–1865.

*details of this work are set out under Technology, below.*

КАТАЛОГЪ библіотеки Ученаго комитета Министерства государственныхъ имуществъ. С.-Петербургъ 1865. pp.[vii]. coll.356. pp.[ii]. lviii. [7000.]

[E. VON LENTHE], Katalog der bibliothek der Hannoverschen landwirtschafts-gesellschaft. Nebst einem vorwort über die einrichtung landwirtschaftlicher bibliotheken. Hannover 1882. pp.xvi.156. [3000.]

D.M.A., Nota de bibliográfica [*sic*] agronómica. Pamplona 1882. pp.54. [738.]

CARL GEROLD, Die literatur der letzten zehn jahre aus dem gesammtgebiete der land- und forstwirtschaft . . . in deutscher, französischer und englischer sprache. Gerold's literarische handbibliothek (vol.ii): Wien [1865]. pp.[iii].239. [6000.]

EDUARD BALDAMUS, Die literarischen erscheinungen der letzten 10 jahre 1856 bis 1865 auf dem gebiete der land- und hauswirthschaft, sowie des gartenbaues. Prag 1867. pp.151. [1750.]

— — 1886–1870. Leipzig 1871. pp.[ii].70. [1250.]

— — 1871–1875. 1876. pp.[ii].95. [1750.]

— — 1876–1880. 1881. pp.[ii].98. [1750.]

— — 1881–1885. 1886. pp.[ii].117. [2000.]

*the title varies.*

ALGEMEINE catalogus. Landbouwhoogeschool: Bibliotheek: Wageningen 1891 &c.

AGRICULTURE. Institute of Jamaica: Classified book-lists (no.1): Kingston 1893. pp.10. [300.]

ACCESSIONS to the department library. Department of agriculture: Library bulletin (no.1 &c.): Washington 1894 &c.

MAX GÜNTZ, Handbuch der landwirtschaftlichen litteratur. Leipzig 1897. pp.[iii].170+xii.312. [1000.]

— — Ergänzungsheft. 1902. pp.[iii].183. [500.]

BARON ARTHUR ADOLF SCHWÄGER VON HOHENBRUCK, Sammlung von landwirtschaftlichen schriften aus der Pariser Weltausstellung 1900. [Wien 1900.] pp.22. [352.]

KATALOG over de Kgl. selskab for Norges vels bibliothek. [Tidsskrift for det norske landbrug:

Bilag:] Christiania 1902. pp.[vi].149. [3300.]

BIBLIOGRAPHIA agronomica universalis. Office international de bibliographie: Bibliographia universalis: Casale Monferrato &c.

i–iv. Dirigé by E. Ottavi et A. Marescalchi. 1903. ff.248.pp.249–264. [2094.]
v–vi. 1904. ff.315. [2838.]
*no more published.*

SUBJECT list of works on agriculture, rural economy, and allied sciences, in the library of the Patent office. Patent office library series (no.15 = Bibliographical series, no.12): 1905. pp.424. [2434.]

BIBLIOGRAPHIE agricole. Revue . . . donnant un compte rendu des mémoires et travaux concernant l'agriculture. N° 1[–3]. Janvier [–juin] 1905. pp.96. [1250.]
*no more published?*

LIST of agricultural books and pamphlets. University of Illinois: College of agriculture: Urbana [1908]. pp.23. [151.]

LIST of books on agriculture and related subjects. University of Massachusetts: [Amherst 1910]. pp. 15. [350.]

LIST of books recommended for grange libraries, with descriptive notes. University of Maine: College of agriculture: The Maine bulletin (vol.xiii, no.1): Augusta 1910. pp.14. [110.]

ONE HUNDRED recent books on agriculture. Carnegie library: Pittsburgh 1910. pp.19. [100.]

CATALOGUE de la bibliothèque. Institut international d'agriculture: Rome 1910. pp.xx.357. [5659.]

— Catalogue systématique. FAO [Food and agriculture organization]: Bureau régional européen: 1948. pp.iii–xx.2747. [75,000.]

— — 1[er] supplément. 1949. pp.[v].36. [600.]

MONTHLY bulletin of the library. Department of agriculture: Washington.

i. 1910. pp.iv.426. [4500.]
ii. 1911. pp.iv.471. [5000.]
iii. 1912. pp.iv.497. [5500.]
iv. (nos.1–6). 1913 (January–June). pp.192. [2000.]

*no more published.*

BULLETIN bibliographique hebdomadaire. Livres et périodiques reçus dans la semaine par la bibliothèque, avec indication pour les périodiques des articles intéressant les divers bureaux de l'Institut. Institut international d'agriculture: Rome.

[i]. 1910. pp.[916]. [10,000.]
ii .1911. pp.2292.68. [14,696.]
iii. 1912. pp.2276.66. [11,569.]
iv. 1913. pp.1240.50. [6665.]
v. 1914. pp.1205.51. [6126.]
vi. 1915. pp.945.44. [4323.]
*no more published.*

HENRY ORMAL SEVERANCE, Books for farmers and farmers' wives. University of Missouri: Bulletin: Library series (no.4): Columbia 1912. pp.24. [150.]

CATALOGUE of the principal official publications relating to agriculture and allied industries filed in the library of the Board of agriculture and fisheries. Board [*afterwards:* Ministry] of agriculture and fisheries: Miscellaneous publications (no.15): 1914. pp.66. [1000.]

A READING list on agriculture. Public library: Special library list (no.9): Kansas City 1915. pp.28. [400.]

FANNIE C. RAWSON, Good books on agriculture, forestry, roads. Kentucky: Library commission: Frankfort [1915]. p.19. [145.]

[GEORGE ERNEST MANWARING], Catalogue of the library of the Royal agricultural society. 1918. pp.[iv].386. [6000.]

## *Agriculture*

THE AGRICULTURAL index. Subject index to a selected list of agricultural periodicals and bulletins. New York.

1916–1918. Edited by Neltje Tannehill Shimer. pp.[iv].1025. [25,000.]

1919–1921. Edited by Florence A. Arnold. pp.lxviii.1226. [30,000.]

1922–1924. Edited by F. A. Arnold and Hazel L. Lewis. pp.lxxx.1100. [25,000.]

1925–1927. pp.cxii.1312. [30,000.]

1928–1930. pp.lix.1504. [40,000.]

1931–1933. pp.lxv.1730. [45,000.]

1934–1936. pp.lxx.1836. [45,000.]

1937–1939. pp.lxviii.1832. [45,000.]

1939–1942. pp.lxxii.2103. [42,500.]

1942–1943. pp.xlix.1537. [30,000.]

1943–1944. pp.xxix.581. [12,500.]

1945–1948. pp.liii.1651. [32,500.]

1948–1951. Edited F. A. Arnold and Arvella C. Malouf. pp.liv.2041. [40,000.]

xiii. 1951–1954. Edited by F. A. Arnold. 1955. pp.xlvi.2146. [50,000.]

xiv. 1954–1956. 1957. pp.xxxviii.1624. [40,000.]

xv. 1956–1958. 1960. pp.xxxvii.1593. [35,000.]

xvi. 1958–1960. 1961. pp.xxxvii.1543. [35,000.]

*in progress; also issued in annual and smaller cumulations.*

HAND-LIST of books on agriculture. National library of Wales: Aberystwyth 1921. pp.16. [400.]
— Third edition. 1926. pp.44. [700.]

УКАЗАТЕЛЬ аграрной литературы (статей и рецензий). Международный аграрный институт: Библиографический отдел: Москва 192?–1930.

[*continued as:*]

ИНДЕКС иностранной сельскохозяйственной литературы ]&c.]. Всесоюзная ассоциация с.-х. библиографии: 1931–?.

AGRICULTURAL library notes. Issued monthly by the United States Department of agriculture library with the cooperation of the libraries of the land grant colleges and the state agricultural experiment stations. Washington.*

i. 1926. Edited by Claribel R[uth] Barnett. ff.(pp.)[iii].157. [1500.]
ii. 1927. ff.(pp.)iii.165. [1500.]
iii. 1928. pp.[ix].167–323. [1500.]
iv. 1929. pp.[vii].325–477. [1500.]
v. 1930. pp.[v].115. [1000.]
vi. 1931. pp.[vii].205. [1500.]

vii. 1932. pp.[v].91. [500.]
viii. 1933. pp.[v].190. [1500.]
ix. 1934. pp.[ix].455. [5000.]
x. 1935. pp.[ix].527. [6000.]
xi. 1936. pp.[iii].615. [7500.]
xii. 1937. pp.[ix].675. [9000.]
xiii. 1938. pp.[ii].vi.709. [10,000.]
xiv. 1939. pp.[ix].650.[10,000.]
xv. 1940. pp.[vii].692. [10,000.]
xvi. 1941. pp.[iv].4.652. [10,000.]
xvii. January–June 1942. pp.[v].270. [4000.]

*no more published.*

[ANNA BEIJER], Katalog över Kungl. landtbruksakademiens bibliotek. Stockholm 1927. pp.viii. 742. [15,000.]

— [—] Tillägg. 1938. pp.341. [7500.]

AGRICULTURE and rural economy. Third Edition. [National book council:] Bibliography (no.58): 1928. pp.[4]. [125.]

— Fifth edition. 1941. pp.8. [200.]

INDEX bibliographique de la question agraire. Institut agraire international: Section de bibliographie: Moscou.

1928. [By P. Kananov and others]. pp.xii.305. [5500.]

1929. pp.xii.235. [3165.]
*no more published.*

AGRICULTURE and connected subjects. Lancashire county library: Bibliography (no.5): [*s.l.*] 1928. pp.17. [275.]

— Supplement. 1930. pp.13. [200.]

A BRIEF summary of the papers published by the staffs of the Department [*afterwards:* School] of agriculture and its associated research institutes. University of Cambridge: School of agriculture: Memoirs (no.1 &c.): [Cambridge].

1928–1929 . . . (no.1). pp.44. [60.]
1929–1930 . . . (no.2). pp.40. [50.]
1930–1931 . . . (no.3). pp.36. [40.]
1931–1932 . . . (no.4). pp.40. [60.]
1932–1933 . . . (no.5). pp.44. [60.]
1933–1934 . . . (no.6). pp.48. [80.]
1934–1935 . . . (no.7). pp.59. [125.]
1935–1936 . . . (no.8). pp.54. [125.]
1936–1937 . . . (no.9). pp.48. [100.]
1937–1938 . . . (no.10). pp.40. [125.]
1938–1940 . . . (nos.11–12). pp.41. [350.]
1940–1942 . . . (nos.13–14). pp.8. [50.]
1942–1945 . . . (nos.15–17). pp.43. [80.]
1945–1946 . . . (no.18). pp.28. [50.]
1946–1947 . . . (no.19). pp.27. [50.]
1947–1948 . . . (no.20). pp.28. [50.]

1948–1949 . . . (no.21). pp.28. [50.]
1949–1950 . . . (no.22). pp.24. [50.]
1950–1951 . . . (no.23). pp.31. [50.]
1951–1952 . . . (no.24). pp.27. [50.]
1952–1953 . . . (no.25). pp.31. [50.]
1953–1954 . . . (no.26). pp.27. [50.]
1954–1955 . . . (no.27). pp.30. [50.]
1955–1956 . . . (no.28). pp.32. [50.]
1956–1957 . . . (no.29). pp.31. [50.]
1957–1958 . . . (no.30). pp.29. [50.]
1958–1959 . . . (no.31). pp.28. [50.]
1959–1960 . . . (no.32). pp.31. [50.]
1960–1961 . . . (no.33). pp.26. [50.]
1961–1962 . . . (no.34). pp.28. [50.]
*in progress.*

ANTON BROSCH, Deutsches schrifttum der agrarpolitik und weltlandwirtschaft von 1920–1929. Reichsministerium für ernährung und landwirtschaft: Bericht über landwirtschaft (n.s., sonderheft 26): Berlin 1930. pp.xii. coll.350. pp.351–374. [7500.]

SZEMLE a bel- és külföldi folyóiratakban, évkönyvekben, önálló művekben megjelent mezőgazdasági érdekű újabb kutatások és kisérletek eredményéről. Mezögazdasági kisérletügyi tanács: Budapest 1932 &c.

AUS dem Internationalen landwirtschaftsinstitut in Rom. . . . Internationale bibliografische umschau. Reichsministerium für ernährung und landwirtschaft: Berichte über landwirtschaft: Sonderabdruck: Berlin.

i. [Von S(igmund) von Frauendorfer]. 1933. pp.190. [2932.]
ii 1934. pp.220. [3118.]
iii. 1935. pp.258. [4499.]
iv. 1936. pp.298. [5173.]
v. 1937. pp.315. [5791.]
vi. 1938. pp.316. [4997.]

*no more published.*

AGRICULTURE (1931). Université de Bordeaux: Institut d'études et de documentation économiques et sociales: Bulletin bibliographique (ser.A, no.13): Bordeaux 1933. ff.25. [700.]

— (1932) . . . (ser.A, no.15): 1933. ff.23. [650.]

BIBLIOGRAPHY of soil science, fertilizers and general agronomy. Imperial bureau of soil science: Harpenden 1931 &c.

*details of this work are entered under Soils, below.*

LISTE des publications. International institute of agriculture: Rome 1936. pp.26. [500.]

— Index décennal des publications . . . (1930–1939). 1941. pp.iv.56. [1000.]

## *Agriculture*

SELECT list of books on agriculture in the science library published since 1920 (including forestry, but excluding ornamental horticulture and veterinary medicine). Science library: Bibliographical series (no.231): 1936. ff.15. [188.]*

— Second edition. . . . (no.469): 1939. ff.14. [219.]*

ВЕСТНИК сельскохозяйственной литературы. Всесоюзная академия сельскохозяйственных наук имени В. И. Ленина [1938–1940: Центральная научная сельскохозяйственная библиотека]: Москва.

1936. pp.176.136.268.250.216. . . . . . . [4000.]

1937. pp.184.172.172.151.155.155.144.140.186.220.159. [4000.]

1938. pp.148.183.179.124.112.156.116.128.116. [4000.]

1939. pp.80.96.96.96.96. . . . . . . [2500.]

1940. pp.80.80.80.64.80.64.72.48.56.80. [2000.]

LA SITUATION de l'agriculture depuis 1933. Université de Bordeaux: Institut d'études et de documentation économiques et sociales: Bulletin bibliographique (no.7): Bordeaux 1937. ff.19. [500.]

POLITIQUE agricole depuis 1933. Université de Bordeaux: Institut d'études et de documentation économiques et sociales: Bulletin bibliographique (no.8): Bordeaux 1937. ff.21. [550.]

READERS' guide to books on agriculture. Library association: County libraries section [no.3]: 1937. pp.[iii].16. [500.]

— [third edition] . . . (new series, no.4): 1949. pp.56. [1000.]

PRODUITS agricoles autres que le blé (depuis 1933). Université de Bordeaux: Institut d'études et de documentation économiques et sociales: Bulletin bibliographique (no.10): Bordeaux 1938. ff.20 [550.]

BIBLIOGRAPHY of agriculture. Department of agriculture: Library [*afterwards:* National agriculture library]: Washington.*

i. 1942. pp.334+106+96+220+81. [12,500.]
ii–iii. 1943. pp.[483]+[674]. [40,000.]
iv–v. 1944. pp.[1217]+[1187]. [54,589.]
vi–vii. 1945. pp.[1112]+[1188]. [57,022]
viii–ix. 1946. pp.[1141]+[1252]. [59,041.]
x–xi. 1947. pp.[1283]+[1604]. [76,990.]
xii. 1948. pp.188+210+255+192+204+185+[146]+145+194+169+171+378. [76,064.]

xiii. 1949. pp.188+215+180+223+218+199+199+201+201+201+156+300. [87,200.]

xiv. 1950. pp.242+204+214+199+231+239+176+214+204+192+464. [87,993.]

xv. 1951. pp.278+203+205+200+218+188+174+192+216+216+182+465. [88,981.]

xvi. 1952. pp.285+214+221+261+229+233+204+213+228+218+170+581. [96,252.]

xvii. 1953. pp.303+231+276+235+277+263+237+178+243+250+158+587. [99,541.]

xviii. 1954. pp.278+267+269+227+227+244+232+207+265+242+179+572. [98.777.]

xix. 1955. pp.277+267+280+213+249+263+244+251+269+245+179+616. [99,940.]

xx. 1956. pp.298+232+239+280+281+256+291+245+269+258+179+652. [98,187.]

xxi. 1957. pp.332+232+243+250+258+268+255+301+314+288+206+671. [98,409.]

xxii. 1958. pp.344+252+280+271+215+

198 + 338 + 247 + 230 + 214 + 179 + 678. [99,470.]

xxiii. 1959. pp.274+220+283+274+253+294+239+209+251+236+173+682. [93,107.]

xxiv. 1960. pp.264+240+289+250+285+266+242+282+259+280+207+728. [96,849.]

xxv. 1961. pp.293+237+265+322+298+304+260+263+251+265+218+762. [94,302.]

xxvi. 1962. pp.282+265+295+299+297+302+310+306+328+318+289+784. [94,968.]

*in progress.*

INFORMATIONAL materials on foreign agriculture, January 1937–June 1943. Department of agriculture: Office of foreign agricultural relations: Division of foreign information and statistics: Washington 1943. pp.[ii].25. [770.]

THE FARMERS' club library. A classified list of books. [Second edition]. 1945. p.44. [600.]

MEDEDEELINGEN van de afdeeling documentatie van het Ministerie van landbouw, visscherij en voedselvoorziening. 's-Gravenhage.

i. 1945.
ii. 1946.
iii. 1947. pp.xxxv.8544.xxxvii.855–1538. [7500.]

[*continued as:*]

Landbouwdocumentatie.

iv. 1948. pp.lii.712+lvii.713–1468. [7402.]
[v]. 1949. pp.lxv.808 + lxxiii.809–1676. [8180.]
[vi]. 1950. pp.lxxx.900 + lxxvi.901–1712. [8307.]
[vii]. 1951. pp.lxiii.772 + xliv.773–1464. [6986.]
[viii]. 1952. pp.xli.728 + xli.729–1496. [6842.]
[ix]. 1953. pp.xli.794+xl.795–1578. [7271.]
[x]. 1954. pp.lviii.850+l.851–1574. [7187.]
[xi]. 1955. pp.xlix.732+lvi.733–1524. [7743.]
xii. 1956. pp.xvi.896 + xxviii.921–1872. [7829.]
xiii. 1957. pp. .912+xlvii.798. [6411.]
xiv. 1958. pp.830+833–1640. [6000.]
xv. 1959.
xvi. 1960. pp.848+849–1680. [6000.]
xvii. 1961. pp.800+801–1654. [8000.]

*in progress.*

BOLETÍN bibliográfico agrícolo. Ministerio de

agricultura: Servicio de capacitación y propaganda: Madrid.

i–iv. 1948. pp.272. [2500.]
vii–x. 1949. pp.224. [2000.]

*no more published?*

AGRARBIBLIOGRAPHIE. . . . Eine zusammenstellung der literatur auf dem gesamtgebiet der land-, forst- und ernährungswirtschaft sowie der grund- und hilfswissenschaften. Hamburg &c.

1945–1946.
1947.
1948.
1949. Herausgegeben: Oswald Asten. pp.xvi. 336. [1870.]
1950. pp.viii.376. [1800.]
1951. pp.viii.360. [1627.]
1952. pp.xvi.392. [2023.]
1953. pp.viii.376. [895.]
1954. pp.viii.376. [1974.]
1955. pp.viii.372. [1896.]
1956. pp.iv.372. [1905.]

*no more published?; limited to writings in german.*

PUBLISHED information on agriculture in foreign countries. Department of agriculture: Office of foreign agricultural relations: Washington 1950. pp.57. [1500.]*

— [another edition]. Published . . . January

1937–December 1953. Foreign agricultural services: 1954. pp.65. [1500.]*
*limited to the agency's own publications.*

BIBLIOGRAPHIE agricole et économique établie... d'après les publications périodiques de langue française conservées (sauf avis contraire) (à la Bibliothèque centrale du) Ministère de l'agriculture. 1950 &c.*
*in progress?*

LISTA de libros y folletos recibidos. Boletin bibliográfico. Ministerio de agricultura y cría: Biblioteca: Caracas 1950 &c.*
*in progress.*

A LIST of publications on the agricultural sciences. Commonwealth agricultural bureaux: Farnham Royal 1952. pp.20. [350.]
— [another edition] Publications list of the Commonwealth agricultural bureaux: 1955. pp. 30. [500.]

BILTEN sadržaja stranih časopisa. Dodatak "Zelenom biltenu". A: Poljoprivreda i šumarstvo. Savezna uprava za unapredenje proizvodnje: [Belgrade].
i. 1952. pp.[329.] [1000.]
*no more published?*

INHALTSVERZEICHNISSE sowjetischer fachzeitschriften in deutscher übersetzung. Reihe IV. Land und forstwirtschaft. Zentralstelle für wissenschaftliche literatur: Berlin.

i. 1952. pp.378. [3500.]
ii. 1953 pp.601. [6500].
iii. 1954. pp.610. [6500.]
iv. 1955. pp.1029. [4000.]
v. 1956. pp.757. [2500.]
vi. 1957. pp.758.65. [2500.]
vii. 1958. pp.792.68. [2500.]
viii. 1959. pp.1106.90. [3000.]
ix. 1960. pp.919.87. [3000.]
x. 1961. pp.1006. [3000.]

*in progress.*

СЕЛЬСКОЕ ХОЗЯЙСТВО. Систематический указатель статей в иностранных журналах. Всесоюзная государственная библиотека иностранной литературы and Центральная научная сельскохозяйственная библиотека: Москва.

1954. pp.[448]. [7500.]
1955. pp.[586]. [9000.]
1956. pp.[894]. [15,000.]
1957. pp.[908]. [15,000.]
1958. pp.[984]. [16,500.]
1959. pp.[864]. [15,000.]

1960. pp.[984]. [16,500.]
1961. pp.[980]. [16,500.]
1962. pp.[1192]. [20,000.]
1963. pp.[1188]. [20,000.]
*in progress.*

G[ALINA] F[EDOROVNA] KRUPNOVA, Что читатъ агроному. Рекомендательный указатель литературы. Государственная . . . публичная библиотека им. М. Е. Сальтыкова-Щедрина. Ленинград 1957. ff.[i].90. [250.]*

HOCHSCHULARBEITEN aus agrarökonomik und agrarsoziologie. Bibliographie abgeschlossener und laufender dissertationen. Forschungsgesellschaft für agrarpolitik und agrarsoziologie (no. 63 &c.): Bonn.*

[i]. 1945–1957. Bearbeitet von Th[eo] Dams. . . . (no.63): 1957. pp.173. [1541.]

[*continued as:*]

Forschungsarbeiten aus agrarökonomik und ländlicher soziologie.

ii. 1957–1960. Bearbeitet von Joachim Lunze. . . . (no.144): 1962. pp.145. [723.]
*in progress.*

J. RICHARD BLANCHARD and HARALD OSTVOLD, Literature of agricultural research. Berkeley &c. 1958. pp.x.231. [1296.]*

## Agriculture

N. A. BANNIKOV and L. M. RASKIN, Сельскохозяйственная литература. Пособие для книжных работников. Книжному работнику о книгах: Москва 1958. pp.244. [2000.]

DAGMAR FRANCISCI, Zemědělské stavby. Universita: Knihovna: Výběravý seznam (no.29): Brně 1958. pp.[v].34. [150.]*

ИНФОРМАЦИОННЫЙ бюллетень новых иностранных книг, поступивших в библиотеку имени В. И. Ленина. Серия II. Биологические, медицинские, сельскохозяйственные науки. Москва 1958 &c.
*in progress.*

CATALOGUS bibliotheek. Ministerie van landbouw en visserij: Cultuurtechnische dienst: Utrecht 1960. ff.[i].viii.100. [1750.]*

FOA library select catalogue of books 1951–1958. Food and agriculture organization of the United nations: [Rome 1961]. pp.xiv.xiv.xiv.700. [12,961.]

PÁLNÉ GYŐRI and LÁSZLÓNÉ MADOS, Agrártudományi egyetemi bibliográfia 1945–1957. Az agrártudományi egyetem központi könyvtárának kiadványai (3rd ser., no.2): Budapest 1961. pp.239. [2758.]

## *Agriculture*

### *Ceylon*

T[OM] PETCH, Bibliography of books and papers relating to agriculture and botany to the end of the year 1915. Peradeniya manuals (vol.iii): Colombo 1925. pp.[iii].256. [5059.]

### *China*

MAO YUNG, Bibliography of chinese literature on agriculture. University of Nanking: Library: Publications (no.1): Nanking 1924. pp.[226]. [3000.]

*the text is in chinese.*

LESLIE T. C. KUO and PETER B. SCHROEDER, Communist chinese monographs in the USDA library. Department of agriculture: Library list (no.71): Washington 1961. pp.87. [800.]*

### *Czechoslovakia*

K OTÁZKÁM budování a upevňování JZD a zvyšování zemědělské výroby. (Bibliografie některých důležitých clánků z českých časopisů, červenec 1956– říjen 1957). Státní knihovna společenských věd (no.108): [Prague] 1957. pp.[iii].27. [300.]*

FRANTIŠEK LAZECKÝ [*and others*], Zemědělske

písemnictví. . . . Soupis českých a slovenských publikací 1945–1955. Československá akademie zemědělských věd [&c.]: Prameny literatury (vol.ii, no.4; iii, nos.1–2): Praha 1957–1958. pp.84+[ii].85–221. [4507.]

VERA KLIMEŠOVÁ, Půda a podnebí. (Příspěvek k rajonisaci zemědělské výroby s přihlédnutím k Moravě). Universita: Knihovna: Výběrový seznam (no.42): Brně 1959. pp.22. [50.]*

VĚRA KLIMEŠOVÁ, Ekonomické stázky JZD. Universita: Výběrobý seznam (no.54): Brně 1960. pp.[ii].28. [60.]*

*Denmark*

K[RISTIAN] HANSEN, Fortegnelse over danske landbrugsskrifter. 18. Danske landmandsforsamling: Odense 1900. pp.xi.159. [3500.]

RECENT literature on danish agriculture in foreign languages. Kgl. veterinær- og landbohøjskole: Bibliotek: Copenhagen 1952. pp.16. [150.]*

OLE KARUP PEDERSEN, VIBEKE AMMUNDSEN and AKSEL MILTHERS, Dansk landbrugs bibliografi. Kgl. danske landhusholdningsselskab: København.

i. Indtil 1814. 1958. pp.vii.461. [6225.]

*in progress.*

# *Agriculture*

## *Dominica*

LUIS FLORÉN LOZANO, La bibliografía agrícola de la republica Dominicana. (Bibliografía preliminar). Reunión técnica de bibliotecarios agrícolas de América latina: Turrialba 1953. ff.[iv].2.52.2.6. [457.]*

## *Egypt*

DRY farming in India, Egypt and South Africa. Science library: Bibliographical series (no.287): 1935 [1937]. single sheet. [22.]*

LIST of publications in English issued by the ministry. Ministry of agriculture: Extension service section: [Cairo] 1937. pp.16. [400.]

## *Eritrea*

CONTRIBUTO ad una bibliografia italiana su Eritrea e Somalia con particolare riferimento all'agricoltura ed argomenti affini. Ministerio degli affari esteri: Istituto agronomico: Firenze [1953]. pp.[v].v.239. [4000.]*

## *Ethiopia*

CONTRIBUTO ad una bibliografia italiana sull'Etiopia con particolare riferimento all'agricoltura ed argomenti affini (1935–1950). Ministero degli affari esteri: Istituto agronomico: Firenze

[1952]. pp.[v].84. [1500.]*

— Contribution to an italian bibliography on Ethiopia 1935–1950 (agriculture and arguments relative to). Translated from italian/P. Gemma. Istituto agronomica per l'Africa italiana: Florence 1952. ff.[i].ii.154. [1250.]*

### *Europe*

INTEGRATION of western Europe: agriculture. Annotated references. Department of state: Division of library and reference services: Bibliography (no.62): [Washington] 1952. ff.13. [100.]*

V. KOREK and JOHANN STARK, Mitteleuropa-bibliographie. Teil I. Agrarfrage, handelspolitik und zusammenschlussbestrebungen in Mitteleuropa. Eine übersicht des schrifttums der jahre 1919 bis 1934. Mitteleuropäische wirtschaftsfragen (n.s. vol.ii): Berlin 1935. pp.xv.265.xvi. [3000.]

LITERATUR zur themengruppe: "landwirtschaft im Sowjetblock". Gesamteuropäisches studienwerk: Bibliothek: Auswahlverzeichnis (no.iv): Vlotho 1957. ff.[iii].45. [750.]*

WERNER SCHLEBITZ, INGE HOFFMANN and baron HELMUT VON VERSCHUER, Die europäische zusammenarbeit auf dem gebiet der landwirtschaft unter besonderer berücksichtigung der tätigkeit internationaler und europäischer organisationen. Deutsche gesellschaft für auswärtige politik: Forschungsinstitut: Aktuelle bibliographien des Europa-archives (no.11): Frankfurt/Main 1957. pp.88. [577.]*

*France*

a. *General*

[A. J. MARCHANT], Répertoire général des livres français sur l'agriculture, l'économie domestique rurale et politique. . . . 1re partie. 1810. pp.xii.36. [591.]

*no more published.*

CATALOGUE des volumes et ouvrages appartenant au Comice agricole de Mamers. Mamers 1861. pp.13. [100.]

[F.] RICHENET, Catalogue de la bibliothèque de la Société d'horticulture et de viticulture de Dole. Dole [printed] 1903. pp.32. [433.]

CATALOGUE de livres d'agriculture. Bibliothèque centrale de prêt de Loir-et-Cher: Blois 1955. pp.[i].20. [350.]*

[MARGUERITE DE CRECY], Bibliographie analytique des enquêtes effectuées par ordre du Ministre du commerce et de l'agriculture de 1800 à 1918.

École pratique des hautes études: Centre de recherches historiques: Histoire des entreprises: 1962. pp.20–76. [68.]

CATALOGUE général des ouvrages d'agriculture. Bibliothèque centrale de prêt des Deux-Sèvres: [Niort 1963]. pp.61. [500.]*

## b. *Colonies*

ÉNUMÉRATION des travaux sur productions et cultures coloniales publiés par le professeur [Frédéric Heim] et ses collaborateurs. [Conservatoire national des arts et métiers:] Chaire d'agriculture et productions agricoles: [1931]. pp.54.

## *Germany*

FRIEDRICH BENEDICT WEBER, Handbuch der ökonomischen literatur; oder systematische anleitung zur kenntnis der deutschen ökonomischen schriften. Berlin 1803. pp.xxxii.375+xx.378.[iv]. [6000.]

— — Erster supplementband. 1809. pp.xxiv. 378.[ii]. [2500.]

— — Zweiter supplementband. Breslau 1816. pp.xxvi.252.[ii]. [2000.]

— — Dritter supplementband. Leipzig 1823. vi.289. [2000.]

— — Vierter supplementband. Breslau 1832. pp.xix.328. [2500.]

— — Fünfter supplementband. Grimma 1842. pp.xiv.377. [2500.]

*the supplements also have the secondary title* Systematisches handbuch der neuesten deutschen ökonomischen literatur.

THEODOR CHRISTIAN FRIEDRICH ENSLIN, Bibliotheca oeconomica, oder verzeichnis aller brauchbaren . . . in Deutschland erschienenen bücher über die land- und hauswirtschaft. Berlin &c. 1825. pp.[ii].143. [3000.]

— — Zweite auflage, von Wilhelm Engelmann. 1841. pp.vi.438. [8000.]

A[NTON] BROSCH, Schrifttum über innere kolonisation. Gesellschaft zur förderung der inneren kolonisation: Berlin 1919. pp.vii.197. [3000.]

A SELECTED list of german publications in the field of agriculture and agricultural economics [since 1939] and political economy [since 1945]. [Omgus (Office of military government): *s.l.* 1948. pp.[ii].26. [291.]*

A SELECTED list of german publications in the fields of agriculture and agricultural economics and political economy. Office of the economic advisor: Food and agriculture group: [Berlin 1949]. pp.24. [301.]

BIBLIOGRAPHIE der landwirtschaftlichen hochschulschriften . . . in deutscher sprache . . . auf dem gebiet der land- und forstwirtschaftswissenschaften. Deutsche bücherei: Sonderbibliographien (no.1): Leipzig.

1945–1952. [Edited by] Werner Dux, Curt Fleischhack. pp.xii.172. [1992.]

1953–1956. pp.xii.339. [3452.]

*in progress.*

*Great Britain*

a. *General*

INDEXES to the subject matters of the reports of the House of commons. . . . Agriculture and corn trade, 1820–1834. 1834. pp.[vi].120. [5000.]

JOHN DONALDSON, Agricultural bibliography: containing a notice of the life and writings of the british authors on agriculture, from the earliest date in 1480 to the present time. 1854. pp.vii [*sic,* viii].138. [531.]

W. FRANK PERKINS, British and irish writers on agriculture. Lymington 1929. pp.[ii].vii.142. [2250.]

— — 3rd edition. 1939. pp.xiii.226. [3000.]

## *Agriculture*

A SELECTED list of publications on agriculture. His majesty's stationery office: 1932. pp.16. [400.]

— [another edition.] For farm and garden. 1939. pp.27. [500.]

SIR E[DWARD] JOHN RUSSELL, Books and the farmer. Association of special libraries and information bureaux: 1933. pp.16. [100.]

A SELECTED and classified list of books in english relating to agriculture, horticulture, etc. in the library of the Ministry of agriculture. Ministry of agriculture and fisheries: Bulletin (no.78): 1935. pp.vi.22. [500.]

— Third edition. 1954. pp.vi.89. [2000.]

H. M. BEDINGTON, Some British books on agriculture, nutrition, forestry and related sciences, 1939–1945. Imperial agricultural bureaux: Joint publication (no.11): [Aberystwyth, Wales] 1946. pp.v.37. [392.]

AGRICULTURE. House of commons: Library: Bibliography (no.18): [1947]. ff.18. [131.]*

A. G. STREET, Farming. National book league: Reader's guide: 1948. pp.16. [200.]

*the bibliography proper is by William Arthur Munford.*

b. *Colonies*

[ULRICH] PHILIP MAYER, Reading list on rural conditions and betterment in the British colonies. Oxford university: Nuffield college: 1947. pp.vii. 121. [1085.]

*Greece*

A[NNE] M[ARIE] HANNAY, Greece. A guide to official statistics of agriculture, population and food supply. Department of agriculture: Bureau of agricultural economics: Agricultural economic bibliography (no.39): Washington 1932. pp.iv. 142. [500.]*

*Hawaii*

ELIZABETH H. LANGDALE, Index to publications of the Hawaii agricultural experiment station. Hawaii agricultural experiment station: Extension bulletin (no.10): Washington 1927. pp.29. [2500.]

*Hungary*

BÉLA KOCZÁNYI, A magyar gazdasági szakirodalom könyvészete, 1860–1888. Szakirodalmi jegyzékek sorozata (vol.i): Kassán 1889. pp.61. [700.]

*printed on one side of the leaf.*

FERENC S. SZABÓ [*and others*], A magyar mezőgazdasági szakirodalom könyvészete (1831–1867).

# *Agriculture*

Bibliographia oeconomica rurali Hungariae (vol. iii): Budapest 1956. pp.660. [3799.]

GYULA WALLESHAUSEN, A mezőgazdasági tudományos kutatás és tájékoztatás segédkönyvei. Az Agrár tudományi egyetem központi könyvtárának kiadványai (4th ser., no.2): Budapest 1957. pp.56. [421.]

*India*

CATALOGUE of books in the library of the Board of revenue (revenue settlement, land records and agriculture), Madras. Madras 1898. pp.180. [4000.]

LIST of references on agricultural machinery in India. Library of Congress: Division of bibliography: [Washington] 1922. pp.6. [75.]*

DRY farming in India, Egypt and South Africa. Science library: Bibliographical series (no.287): 1935 [1937]. single sheet. [22.]*

*Indochina*

PIERRE CARTON, Catalogue de la bibliothèque. Institut scientifique de l'Indochine: Hanoï 1922. pp.xi.433. [5000.]

*Indonesia*

CATALOGUS van de bibliotheek van het algemeen proefstation der A. V. R. O. S. [Algemeene vereeniging van rubberplanters ter oostkust van Sumatra] te Medan. Kampong Baru 1957–1958. pp.[ii].259+[ii].615. [4000.]*

— Eerste aanvulling. 1958. pp.63.45. [750.]*

*Italy*

MARCO [ANTONIO] LASTRI, Biblioteca georgica, ossia catalogo ragionato degli scrittori di agricoltura, veterinaria, agrimensura, meteorologia, economia pubblica, caccia, pesca, ec. spettanti all' Italia. Firenze 1787. pp.xii.152. [1000.]

[GIUSEPPE MORETTI], Bibliografia agronomica, saggio di un catalogo ragionato de' libri d'agricoltura e veterinaria scritti in italiano o all' Italia spettani. Milano 1844. pp.iii.328. [1250.]

CATALOGO delle pubblicazioni e dei prodotti agrari presentati dalla Direzione generale dell'agricoltura. Ministerio di agricoltura, industria e commercio: Roma 1890. pp.179. [2000.]

VITTORIO NICCOLI, Saggio storico e bibliografico dell' agricoltura italiana, dalle origini al 1900. Nuova enciclopedia agraria italiana (vol.i): Torino 1902. pp.575. [10,000.]

BIBLIOGRAFIA scientifico-tecnica italiana, 1928. Gruppo IX. Agricoltura. Bologna [1929]. pp.[vi]. 281. [6799.]

## *Agriculture*

BIBLIOGRAFIA italiana—Gruppo D. Agricoltura. Consiglio nazionale delle ricerche: Bologna [Piacenza, Roma].

1929. pp.[v].148. [2219.]
1930. pp.[v].139. [1725.]
1931. pp.[v].144. [2036.]
1932. pp.[v].155. [2186.]
1933. pp.[v].168. [2380.]

## *Java*

LIJST van publicaties van het proefstation midden- en oost-Java . . . 1911 tot 1936. De Bergcultures (vol.x, no.26: supplement): Batavia 1936. coll.16.

## *Kazakhstan*

F. V. ZHIZNEVSKY, Опыт передовиков сельского хозяйства Казахской ССР. Аннотированный указатель литературы. Академия наук Казахской ССР: Центральная научная библиотека: Алма-Ата 1955. pp. 291. [750.]

## *Libya*

CONTRIBUTO ad una bibliografia italiana sulla Libia con particolare riferimento all'agricoltura ed argomenti affini. Ministero degli affari esteri: Istituto agronomico: Firenze [1953]. pp.[v].vii.193. [3500.]*

### *Mexico*

LISTA de las publicaciones disponibles. Secretaria de agricultura y fomento: México 1921. pp.15. [275.]

ALGUNAS publicaciones de agrónomos. Sociedad agronómica mexicana: Biblioteca de la Feria de libro: México 1943. ff.[11]. [100.]*

LUIS CASTILLA LEDÓN, RITA MARTÍNEZ and GABRIEL SALDÍVAR, Bibliografía agrícola y agraria de México. [on cover: Tomo I]. México 1946. pp.3–525. [3047.]
*no more published.*

ÍNDICE bibliográfico del Centro de investigación y de estudios avanzados. Sección 5ª .... II. Agricultura, zootécnica e industrias de la alimentación, Instituto politécnico nacional: México 1962 &c.
*in progress.*

### *Moravia*

SPISY vyšlé z Moravského zemského výzkumného ústavu zemědělského v Brně v letech 1899–1926. Brno [1927]. pp.36. [375.]

### *Netherlands*

C[ORNELIS] R[UDOLPHUS] HERMANS, Beredeneerd overzigt der landbouwkundige schriften betrekkelijk de provincie Noord-Braband. Provinciaal genootschap van kunsten en wetenschappen in Noord-Braband: 's-Hertogenbosch 1845. pp.60. [50.]

[WINAND CAREL HUGO STARING], Alphabetische naamlijst van alle nederlandsche geschriften, over landbouw, veeteelt, vee-artsenijkunde, houtteelt en tuinbouw, zooals die afzonderlijk uitgegeven, of in tijdschriften en werken van genootschappen opgenommen zijn Utrecht 1851. pp.[ix].103. [1000.]

BOEKEN, tijdschriften en platen over landbouw, veeteelt, paarden, zuivelbereiding, tuinbouw enz. bijeengebracht voor de tentoonstelling gehouden ter gelegenheid van het 60-jarig bestaan van de Friesche maatschappij van landbouw. Leeuwarden 1912. pp.33.

LIJST va opstellen, voorkomende in nos. I–XXXVI en 37 der Verslagen van landbouwkundige onderzoekingen der rijkslandbouwproefstations. [*s.l.* 1932]. pp.18.

ACTA agralia vadensia. A record of publications on agriculture, mostly published under the auspices of the Landbouwhoogeschool. Wageningen.

i. 1934. Compiled by H[avik] N[icolaas] Kooiman and H[endrik] J[acobus] Vene-

ma. 1936. pp.[iv].55. [100.]*
*no more published.*

*Peru*

JOSÉ G. OTERO, Guía bibliográfica consultiva de obras, folletos, revistas y publicaciones, sobre temas relacionados con la agricultura y la ganadería nacionales. Ministerio de fomento: Lima 1906. pp.99. [750.]

*Philippines*

EDUARDO R. ALVARADO, Philippine agricultural bibliography. Part I. Check list of books, circulars, and miscellaneous publications of the Bureau of agriculture. Bureau of agriculture: Manila 1926. pp.38. [350.]
*no more published.*

BASILIO HERNANDEZ, Philippine bibliography of the nine major crops of the Philippines: rice, sugar cane, abaca, coconut, tobacco, corn, maguey, coffee and cacao. Bureau of science library: Manila, P.I. 1933. ff.[i].ii. pp.132. [1500.]*

BASILIO HERNANDEZ and LAUREANA ESTRELLA [VILLANUEVA], Philippine bibliography of the minor crops of the Philippines. Department of agriculture and commerce: Bureau of science: Manila 1934. ff.[iii].ii.[i].166.iii. [2000.]

## *Agriculture*

LIST of available publications. Department of agriculture and commerce: Manila 1936. pp.41. [400.]

### *Poland*

ZDZISŁAW KOSIEK, Bibliografia polskich bibliografii gospodarstwa wiejskiego. Warszawa 1962. pp.xvi.119. [1036.]

---

M[ARIAN] GÓRSKI, Zarys historji gleboznawstwa w Polsce (z bibljografią od roku 1919). Poznań 1927. pp.14. [75.]

[JÓZEF BOROWIK and STEFAN KOPEĆ], Bibliografja Pulawska, 1918–1927. Spis publikacyj pracowników Pánstwowego instytutu naukowego gospodarstwa wiejskiego ogloszonych w latach 1918–1917. Bydgoszcz 1930. pp.59. [609.]

BIBLIOGRAFIA polskiego pišmiennictwa rolniczego za lata 1945–1953. Centralny instytut rolniczy: Warszawa 1954. pp.xxi.496. [15,479.]

ZDZISŁAW KOSIEK, Bibliografia polskich bibliografii gospodarstwa wiejskiego. Warszawa 1962. pp.xvi.119. [1036.]

### *Portugal*

[PEDRO WENCESLAU DE] BRITO ARANHA, Exposição agricola de 1884 na real tapada da Ajuda.

*Agriculture*

Instrucção agricola . . . bibliographia. Lisboa 1884. pp.46. [439.]

CATÁLOGO das publicações da extinta Estação agrária central. Ministério da agricultura: Direcção geral dos serviços agrícolas: Lisboa 1937. pp.15. [53.]

*Rhodesia*

ELIZABETH H. BERNSTEIN, Farming in Southern Rhodesia. A bibliography. University of Cape Town: School of librarianship: Bibliographical series: [Capetown] 1950. ff.30. [200.]*

*Romania*

A[NNIE] M[URRAY] HANNAY, Rumania. A guide to official statistics of agriculture, population and food supply. Department of agriculture: Bureau of agricultural economics: Agricultural economic bibliography (no.49): Washington 1933. pp.[ii]. iv.220. [2000.]*

*Russia*

N[IKOLAI] M[IKHAILOVICH] MIKHEEV, Библиографические указатели сельскохозяйственной литературы, 1783–1954 годы. Центральная научная сельскохозяйственная библиотека [&c.]: Москва 1956. pp.192. [608.]

[VLADIMIR IZMAILOVICH MEZHOV], Библіографія вопроса объ улучшеніи быта помѣщичьихъ крестьянъ въ южнорусскомъ краѣ, съ 1857–1860 г. Санктпетербургъ. 1861. pp.16. [75.]

V. I. MEZHOV, Крестьянскій вопросъ въ Россіи. Полное собраніе матеріаловъ для исторіи крестьянскаго вопроса на языкахъ русскомъ и иностранныхъ напечатанныхъ въ Россіи и заграницею 1764–1864. Санктпетербургъ. 1865. pp.[iii].viii.[v].422. [3305.]

УКАЗАТЕЛЬ хронологическій, предметный и алфавитный выданныхъ в Россіи привилегій по сельскохозяйственной части съ 1843 по 1868 годъ. С.-Петербургъ 1868. pp.18. [86.]

V. I. MEZHOV, Земскій и крестьянскій вопросы. Библіографическій указатель книгъ и статей, вышедшихъ по первому вопросу съ самаго начала введенія въ дѣйствіе земскихъ учрежденій и рапѣс, по второму — съ 1865 г. вплоть до 1871 г. С.-Петербургъ 1873. pp.[ii].80. [1500.]

I[VAN] I[VANOVICH] MAMONTOV, Указатель изданій министерства земледѣлія и государственныхъ имуıпествъ по сельскотозяйственной и лѣсной части ... (1894–1900). С.-Петербургъ 1901. pp.xii.831. [346.]

— — за 1901 годъ. 1903. pp.[ii].iv.272. [129.]

G. I. USPENSK, Краткій указатель литературы по крестьянскому вопросу. Тула 1904. pp.15. [224.]

A. D. PEDASHENKO, Указатель книгъ, журнальныхъ и газетныхъ статей по сельскому хозяйству за 1910 годъ. Главное управленіе землеустройства и земледѣлія: Отдѣлъ сельской экономіи и сельскохозяйственной статистики: С.-Петербургъ 1913. pp.xv.480. [12,500.]

СИСТЕМАТИЧЕСКИЙ каталог изданий Народного комиссариата земледелия 1919–21 гг. Москва 1921. pp.32. [600.]

YA[KOV] P[ETROVICH] OGUS, Список изданий по сельскому хозяйству, вышедших в 1921 году. Воронеж 1923. pp.16. [477.]

N[IKOLAI] M[IKHAILOVICH] MIKHEEV. Указатель основной литературы по сельскому хозяйству. Всероссийский кооперативный издательский союз 'Книгосоюз': Москва 1929. pp.131. [2000.]

## *Agriculture*

АГРАРНАЯ [1931–1934: Сельскохозяйственная] литература СССР. Международный аграрный институт: Библиографический отдел: Москва.

1930. Под редакцией И. Н. Кобленца и И. И. Полосина. coll.112.90.90. 158. 252. [8000.]

1931. Под редакцией И. Н. Ковленца. coll.304.216.280. [8000.]

1932. pp.xvi.coll.284.472.406.414. [16,000.]

1933. pp.xii.coll.256.368.344.336.436.420. [22,500.]

1934. pp.xvi.coll.232.280.392.272.514.448. 348. [25,000.]

*no more published?*

[L. A. PLAKHOV], Сельско-хозяйственная литература. Москва 1940. pp.74. [400.]

СЕЛЬСКОХОЗЯЙСТВЕННАЯ литература СССР. Систематический указатель. Центральная научная сельскохозяйственная библиотека: Москва.

1951. pp.239.247.249.315.281.245.263.334. [50,000.]

1952.

1953.

1954.

1955.

1956. pp.148.180.216.208.244. [50,000.]
1957. pp.263.268.270.347.268.347.385.240. [50,000.]
1958. pp.256.364.380.334.288.280.267.264. 308.344. [50,000.]
1959. pp.284.327.394.250.248.252.255.242. 282.268.251.191.341. [60,000.]
1960. pp.183.222.260.231.238.299.216.191. 223.239.195.195.319. [60,000.]
1961. pp.219.180.228.247.212.159.224.233.
1962. pp.215.216.172.175.183.263.315.232. 223.240.287.304. [48,604.]
1963. pp.231.320.324.276.295.232.280.251. 282.292. . .

*in progress.*

INHALTSVERZEICHNISSE sowjetischer fachzeitschriften in deutscher übersetzung. Reihe IV. Land- und forstwirtschaft. Zentralstelle für wissenschaftliche literatur: Berlin.

i. 1952. pp.378. [2000.]
ii. 1953. pp.600. [3000.]
iii. 1954. pp.610. [3000.]
iv. 1955. pp.1029. [4000.]
v. 1956. pp.757. [2500.]
vi. 1957. pp.758.65. [2500.]
vii. 1958. pp.792.68. [2500.]
viii. 1959. pp.1106.90. [3000.]

ix. 1960. pp.919.87. [2750.]
x. 1961. pp.1062.91. [3000.]
xi. 1962. pp.1041. [3000.]

*in progress.*

V[ERA] V[LADIMIROVNA] ANUCHEVA, Квадратный и квадратно-гнездовой способ возделывания сельскохозяйственных культур. Рекомендательный список литературы. Государственная . . . публичная библиотека имени М. Е. Салтыкова-Щедрина: Ленинград 1954. pp.19. [50.]

ПРЕДВАРИТЕЛЬНЫЙ каталог по сельскому хозяйству. Институт по изучению истории и культуры СССР: Сообщения библиотеки института (no.4): Мюнхен 1954. pp.42. [400.]

СОВЕТСКИЙ народ в борьбе за крутой подъем сельского хозяйства СССР. Библиографический указатель. Академия наук СССР: Фундаментальная библиотека общественных наук: Москва.*

i.
ii. 1955–1956. pp.[viii]. 327. [3436.]

L. M. INKOVA, Наука и техника в сельском хозяйстве. Краткий рекомендательный указатель. Государственная . . . библиотека СССР имени В. И. Ленина: Библиографический отдел: В помощь лектору: Москва 1955. pp.88. [700.]

A. P. MILYAEV, Основы сельского хозяйства. Программа в помощь самообразованию работников сельских культурно-просветительных учреждений. Государственная . . . публичная библиотека СССР имени В. И. Ленина: Москва 1955. pp.64. [250.]

[M. S. VOINOV and N. M. METELKIN], Пропаганда сельскохозяйственной литературы в районных и сельских библиотеках. Государственная ордена Ленина библиотека СССР имени В. И. Ленина: Москва 1955. pp.95. [200.]

РАЗВИТИЕ сельского хозяйства СССР в шестой пятилетке (1956–1960). Краткий рекомендательный список литературы. Государственная . . . публичная библиотека имени М. Е. Салтыкова-Щедрина: Ленинград 1956. pp.7. [25.]

[A. S. MUILNIKOV, *ed.*], Развитие сельского хозяйства СССР в шестой пятилетке (1956–1960). Краткий рекомендательный список литературы. Публичная библиотека имени М. Е. Салтыкова-Щедрина: Ленинград 1956. pp.7. [30.]

V[ALENTINA] E[VGENEVNA] SHABANOVA, Наука и техника на полях нашей страны. Беседа о книгах. Государственная... библиотека СССР имени В. И. Ленина: За новые успехи в шестой пятилетке: Москва 1956. pp.20. [16.]

G[ALINA] F[EDOROVNA] KRUPNOVA, Путеводитель по сельскохозяйственным справочникам. Государственная ... библиотека имени М. Е. Салтыкова-Щедрина: Ленинград 1958. pp.124. [314.]

E. V. BAZHANOVA, Проблемы производительности труда и себестоимости продукций в сельском хозяйстве СССР после Великой отечественной войны. Библиографический указатель книг и журнальных статей за 1945–1957. Академия наук СССР: Фундаментальная библиотека общественных наук: Москва 1959. pp.160. [1274.]

N. M. METELKIN, Помощь библиотек сельскохозяйственному производству. Сборник материалов. Государственная ... библиотека СССР имени В. И. Ленина: Научно-методический отдел библиотековедения и библиографии: Москва 1955. pp.99. [250.]

РАСТЕНИЕВОДСТВО. Рекомендательный указатель литературы. В помощь организаторам сельскохозяйственного производства. Государственная ордена Ленина библиотека СССР имени В. И. Ленина: Москва 1960. pp.247. [1500.]

F. S. GINZBURG and E. V. PARKHOMENKO, В помощь агроному на производстве. Аннотированный указатель литературы. Центральная научная сельскохосяйственная библиотека: Москва 1960. pp.144. [500.]

E[LENA] N[IKOLAEVNA] MORACHEVSKAYA, Библиография по районированию и размещению сельского хозяйства СССР (1818–1960). Академия наук СССР: Совет по изучению производительных сил: Сектор сети специальных библиотек: Москва 1961. pp.200. [2000.]

*Slovenia*

JANKO ŠLEBINGER and JANEZ MARENTIČ, Bibliografija slovenske kmetijske literature v letih 1919–1938. Izdanja Kmetijske zbornice Dravske banovine (vol.2): Ljubljani 1939. pp.[xii].207. [4000.]

## *Agriculture*

### *Somalia*

CONTRIBUTO ad una bibliografia italiana su Eritrea e Somalia con particolare riferimento all'agricoltura ed argomenti affini. Ministerio degli affari esteri: Istituto agronomico: Firenze [1953]. pp.[v].v.239. [4000.]*

### *Spain*

[BERNARDO RODRIGUEZ], Catálogo de algunos auctores españoles que han escrito de veterinaria, de equitación, y de agricultura. Madrid 1790. pp.87. [400.]

BRÁULIO ANTÓN RAMÍREZ, Diccionario de bibliografía agronómica y de toda clase de escritos relacionados con la agricultura. Madrid 1865. pp.xix. 1018. [2375.]

WILLIAM E[RNEST] CROSS, La estación experimental agrícola de Tucumán de 1914 a 1946. Trabajos e informes publicados. Barcelona &c. 1952. pp.xxiv.715. [900.]

FRANCISCO TOLSADA PICAZO, Bibliografía española de agricultura (1495–1900). Instituto nacional del libro español: Madrid 1953. pp.iii–lxvii.125. [2500.]

AMANDO MELÓN Y RUIZ DE GORDEJUELA, Guia

bibliográfica para una geografia agraria de España. Instituto "Juan Sebastian Elcano" de geografía: Madrid 1961. pp.iii–xxxvi.887. [6375.]

*Sudan*

R[OBERT] L[ANIER] KNIGHT and B. M. BOYNS, Agricultural science in the Sudan. A bibliography with abstracts. Arbroath 1950. pp.251. [969.]

LIST of publications. Ministry of agriculture: Agricultural publications committee: Khartoun 1950 &c.*

*Sweden*

DEN SVENSKA lantbrukslitteraturen. Bibliografisk förteckning. Kungl. lantbruksakademie: Uppsala.

i. Från äldsta tid t. o. m. år 1800. Uppgjord av P[er] M[agnus] Hebbe. 1939. pp.xvi.333. [2194.]

ii. 1801–1850. Utgiven av C. Björkbom. 1945. pp.xvi.372. [2970.]

*Switzerland*

F. and E[RNST] ANDEREGG, Landwirtschaft. Centralkommission für schweizerische landeskunde: Bibliographie der schweizerischen landeskunde (Section V.9a–b.i–vi): Bern 1894–1895. pp.xii.

246+x.247–460+viii.461–616–+vi.617–656+ix. 657–856+vi.857–971. [17,500.]

*Tropics*

J[OHN] C[HRISTOPHER] WILLIS, Literature of tropical economic botany and agriculture. II. 1904–08 to 1907–10. [Colombo 1911]. pp.45. [1500.]

SELECTED list of publications on tropical agriculture. Pan american union: Washington 1930. ff.[i].90. [1000.]*

BIBLIOGRAPHY of tropical agriculture. International institute of agriculture: Rome.

1931. pp.[v].70. [400.]
1932. pp.iii–viii.88. [400.]
1933. pp.iii–viii.140. [750.]
1934. pp.vii.247. [1000.]
1935. pp.vii.256. [1250.]
1936. pp.vii.327. [1500.]
1937. pp.vii.420. [1750.]
1938. pp.vii.466. [2000.]
1939. pp.iii.497. [1500.]
1940. pp.iii.302. [1000.]
1941–1942. pp.iii.365. [1000.]

*no more published.*

PERIODICALS and books relating to tropical agriculture and forestry. Science library: Bibliographical series (no.75): [1933]. ff.[i].6. [98.]*

### *Tunisia*

LISTE des publications disponibles. Service botanique et agronomique de Tunisie. Ariana [1951]. pp.24. [400.]

— Supplément. [1954]. pp.4. [50.]

### *United States*

#### a. *General*

A GENERAL index of the agricultural reports of the Patent office, for the twenty-five years, from 1837 to 1861; and of the Department of agriculture, for fifteen years, from 1862 to 1876. [Department of agriculture: Reports (no.14):] Washington 1879. pp.225. [10,000.]

— Supplement . . . for the years 1877 to 1885 . . . [no.42]: 1886. pp.113. [5000.]

T. H. MCKEE, Reports of the Committee on agriculture, House of representatives, from the organization of the committee, May 3, 1820, to the close of the forty-ninth congress, 1887, inclusive. Washington 1887. pp.7. [175.]

T. H. MCKEE, Reports of the Committee on agriculture and forestry, United States Senate, from the organization of the committee, March 4,

1863, to the close of the forty-ninth congress, 1887. Washington 1887. pp.6. [40.]

DIGEST of the annual reports of the agricultural experiment stations in the United States. Department of agriculture: Office of experiment stations (no.2): Washington.

1888. pp.258+173. [1000.]

[*continued as:*]

Experiment station record. Department of agriculture: Office of experiment stations: Washington.

i. 1889–1890. pp.iii.358. [1000.]
ii. 1890–1891. pp.xxxii.853. [2500.]
iii. 1891–1892. pp.xxxviii.1059. [3000.]
iv. 1892–1893. pp.xxx.1086. [3000.]
v. 1893–1894. pp.xxxiv.1227. [4000.]
vi. 1894–1895. pp.xii.1134. [4000.]
vii. 1895–1896. pp.xii.1092. [5000.]
viii. 1896–1897. pp.xiv.1128. [5000.]
ix. 1897–1898. pp.xvi.1198. [5000.]
x. 1898–1899. pp.xvi.1208. [5000.]
xi. 1899–1900. pp.xvi.1208. [5000.]
xii. 1900–1901. pp.xvi.1189. [5000.]
— General index. . . . Vols. I to XII . . . and to Experiment station bulletin no.2. 1903. pp.671.

xiii. 1901–1902. pp.xviii.1206. [5000.]
xiv. 1902–1903. pp.xviii.1249. [5000.]
xv. 1903–1904. pp.xviii.1247. [5000.]
xvi. 1904–1905. pp.xviii.1255. [5000.]
xvii. 1905–1906. pp.xx.1317. [5000.]
xviii. 1906–1907. pp.xx.1297. [5000.]
xix. 1907–1908. pp.xx.1316. [5000.]
xx. 1908–1909. pp.xviii.1341. [5000.]
xxi. 1909–1910. pp.xiii.911. [3000.]
xxii. 1910. pp.xiii.900. [3000.]
xxiii.1910–1911. pp.xiv. 918. [3000.]
xxiv. 1911. pp.xiii.898. [3000.]
xxv. 1911–1912. pp.xvi.1003. [4000.]
— General index. . . . Vols. XIII to XXV . . . by M. D. Moore and William Henry. 1913. pp.1159.
xxvi. 1912. pp.xiii.1009. [4000.]
xxvii. 1912–1913. pp.xv.1004. [4000.]
xxviii. 1913. pp.xv.1004. [4000.]
xxix. 1913–1914. pp.xvi.1004. [4000.]
xxx. 1914. pp.xii.1000. [4000.]
xxxi. 1914–1915. pp.xv.1011. [4000.]
xxxii. 1915. pp.xix.1014. [4000.]
xxxiii. 1915–1916. pp.xxi.1010. [4000.]
xxxiv. 1916. pp.xxvi.1010. [4000.]
xxxv. 1916–1917. pp.xxiii.996. [4000.]
xxxvi. 1917. pp.xxv.988. [4000.]

xxxvii. 1917–1918. pp.xxv.990. [4000.]

xxxviii. 1918. pp.xxvii.1001. [4000.]
xxxix. 1918–1919. pp.xxix.1002. [4000.]
xl–xli. 1919. pp.xxvii.1004+xxiv.1007. [7500.]
xlii–xliii. 1920. pp.xxiii.1007+xxiv.1006. [7500.]
xliv.–xlv. 1921. pp.xxvi.998+xxvii.995. [7500.]
xlvi–xlvii. 1922. pp.xxii.996+xviii.1001. [7500.]
xlviii–xlix. 1923. pp.xvi.992+xviii.1004. [7500.]
l. 1924. pp.xvi.992. [4000.]

— General index. . . . Vols. XXVI to L . . . by Martha C. Gundlach. 1926. pp.709.
li. 1924. pp.xvii.992. [4000.]
lii–liii. 1925. pp.xix.999+xvii.1001. [7500.]
liv–lv. 1926. pp.xvii.998+xxi.1002. [7500.]
lvi–lvii. 1927.pp.xvii.1001+xvii.1002.[7500.]
lviii–lix. 1928. pp.xviii.1000+xvii.1001. [7500.]
lx. 1929. pp.xvi.1005. [4000.]

— General index. . . . Volumes 51 to 60. 1932. pp.v.677.
lxi. 1929. pp.xxii.1005. [4000.]

lxii–lxiii. 1930. pp.xxii.1000+xxii.998. [7500.]

lxiv–lxv. 1931. pp.xix.1000+xx.1008. [7500.]

lxvi–lxvii. 1932. pp.xxi.1009+xix.895. [7500.]

lxviii–lxix. 1933. pp.xviii.991+xx.1044. [7500.]

lxx. 1934. pp.xxiii.1012. [4000.]

— General index. . . . Volumes 61 to 70. 1937. pp.iv.752.

lxxi. 1934. pp.xx.1020. [4000.]

lxxii–lxxiii. 1935. pp.xxi.1031+xxi.1029. [7500.]

lxxiv–lxxv. 1936. pp.xx.1029+xxiii.1046. [7500.]

lxxvi–lxxvii. 1937. pp.xviii.1033+xxii. 1038. [7500.]

lxxviii–lxxix. 1938. pp.xxii.1022+xx.842. [7500.]

lxxx–lxxxi. 1939. pp.xxii.1002+xxi.1054. [7500.]

lxxxii–lxxxiii. 1940. pp.xx.1013+xxi.1019. [7500.]

lxxxiv–lxxxv. 1941. pp.xii.1019+xii.1016. [7500.]

lxxxvi–lxxxvii. 1942. pp.xii.1020+xv.1056. [7500.]
lxxxviii–lxxxix. 1943. pp.xiii.1011+xiv.885. [7500.]
xc–xci. 1944. pp.xiii.981+xii.901. [7500.]
xcii–xciii. 1945. pp.xiii.1010+xiii.941. [7500.]
xciv–xcv. 1946. pp.xi.975+xii.1033. [7500.]
— General index. . . . Volumes 71 to 80. 1949. pp.iv.832.

*no more published.*

PUBLICATIONS issued by the U. S. Department of agriculture [*now after minor changes:* Monthly list of publications]. Washington 1892 &c.

*in progress; these lists are cumulated and indexed in various forms (see also the entry under 1879 above)*:

— Geo. F. Thompson, Index to authors with titles of their publications appearing in the documents of the U. S. Department of agriculture, 1841 to 1897. Department of agriculture. Bulletin (no.4): Washington 1898. pp.303. [1500.]

— R. B. Handy and Minna A. Cannon, List by titles of publications of the United States Department of agriculture from 1840 to June 1909, inclusive . . . (no.6): 1902. pp.216. [5000.]

— List of publications of the Agriculture

department, 1862–1902. Bibliography of United States public documents: Department lists (no.1): Washington 1904. pp.623. [20,000.]

— Doris Stockdale, Publications of the U. S. Department of agriculture. Department of agriculture: Miscellaneous circular (no.103): Washington 1927. pp.110. [5000.]

— Mabel G[raybill] Hunt, List of publications of the United States Department of agriculture from January, 1901, to December, 1925. [Department of agriculture:] Miscellaneous publication (no.9): Washington 1927. pp.v.182. [7500.]

— — 1926–1930 . . . (no.153): 1932. pp.ii.46. [2000.]

— — 1931–1935 . . . (no.252): 1936. pp.ii.64. [2750.]

— — 1936–1940. Compiled . . . by Mabel Hunt Doyle. Department of agriculture: Miscellaneous publication (no.443): 1941. pp.[ii].68. [2500.]

— — 1941–1945. [By] Bertha Louise Zoeller . . . (no.611):

— Mary A. Bradley, Index to publications of the United States Department of agriculture, 1901–1925. Washington 1932. pp.vi.2689. [250,000.]

— — 1926–1930. 1935. pp.v.694. [60,000.]

— Numerical list of current publications of the

# *Agriculture*

United States department of agriculture. By Fred L[yon] Zimmerman and Phyllis R[ogers] Read. Department of agriculture: Miscellaneous publication (no.450): Washington 1941. pp.v.929. [20,000.]

— Eleanor W. Clay, List of available publications of the United States Department of agriculture. [Department of agriculture:] Miscellaneous publications (no.60): Washington [1948]. pp.iv.187. [4000.]

— — Revised edition. 1951. pp.iv.173. [4000.]

— — [another edition]. Compiled by E.W. Clay and Ella J. Green. Department of agriculture: Publications division: List (no.11): 1962. pp.ii.104. [3500.]

*there are several intermediate editions.*

LIST of [agricultural experiment] station publications received by the Office of experiment stations. Department of agriculture: Office of experiment stations: Washington 1906–1943.

*454 bi-monthly, then monthly numbers issued, with slight variations in title, in various forms and by several agencies in succession; continued in the* Bibliography of agriculture *of the Department of agriculture.*

LIST of references on the settlement of soldiers

and sailors on the land. Library of Congress: Washington 1918. ff.5. [62.]*

— [another edition]. 1919. ff.11. [145.]*

LIST of references on farm labor in the United States (with special reference to conditions of life and labor of farm workers). Library of Congress: [Washington] 1921. ff.10. [142.]*

LIST of references on blocs, with special reference to the agricultural bloc in Congress. Library of Congress: [Washington] 1922. ff.6. [45.]*

A SHORT list of references on the stabilization of prices of agricultural products in the United States. Library of Congress: Washington 1923. ff.6. [53.]*

A LIST of references on the present agricultural situation in the United States. Library of Congress: Division of bibliography: [Washington] 1924. ff.17. [238.]*

[E. LUCY OGDEN and MARTHA L. GERICKE], List of bulletins of the agricultural experiment stations in the United States from their establishment to the end of 1920. Prepared in the Office of experiment stations. Department of agriculture: Department bulletin (no.1199): Washington 1924. pp. 185. [12,500.]

— — Supplement.

i. 1921–1922. By Catherine E[mma] Pennington. 1924. pp.24. [750.]

ii. 1923–1924. 1926. pp.54. [1000.]

iii. 1925–1926. 1927. pp.62. [1000.]

[iv]. 1927–1928. Department of agriculture: Miscellaneous publication (no.65): 1930. pp.77. [1250.]

[v]. 1929–1930. . . . (no.128): 1932. pp.88. [1250.]

[vi]. 1931–1932. . . . (no.181): 1934. pp.77. [1250.]

[vii]. 1933–1934. . . . (no.232): 1936. pp.81. [1250.]

[viii]. 1935–1936. . . . (no.294): 1938.

[ix]. 1937–1938. . . . (no.362): 1940.

[x]. 1939–1940. . . . (no.459): 1942.

[xi]. 1941–1942. Compiled by Helen V. Barnes. Department of agriculture: Bibliographical bulletin (no.4): 1944. pp.70. [1000.]

MARY G[OODWIN] LACY, Long-time agricultural programs in the United States—national, regional, and state. Department of agriculture: Bureau of agricultural economics: Agricultural economics bibliography (no.5): Washington 1925. ff.[ii].20. [100.]*

LIST of recent references on federal agricultura relief. Library of Congress: Division of bibliography: [Washington] 1926. pp.13. [172.]*

LIST of speeches, etc., on federal agricultural relief as printed in the Congressional record, 68th Congress, first session, v.67, current file. Library of Congress: Division of bibliography. [Washington] 1926. pp.16. [276.]*

CORA L[UCINDA] FELDKAMP, Agricultural and home economics extension in the United States. A selected list of references. Department of agriculture: Bibliographical contributions (no.18): Washington 1928. ff.[v].56. [600.]*

LOUISE O[LDHAM] BERCAW, Labor requirements of farm products in the United States . . . since 1922. Department of agriculture: Bureau of agricultural economics: Agricultural economic bibliography (no.26): Washington 1929. pp.[ii].62. [214.]*

ESTHER M[ARIE] COLVIN, Group and chain farming in the United States, January 1930–March 1933. With some references to group farming in foreign countries. Department of agriculture:

## *Agriculture*

Bureau of economic bibliography: Agricultural economic bibliography (no.46); Washington 1933. pp.[iii].32. [100.]*

LOUISE O[LDHAM] BERCAW, The domestic allotment plans for the relief of agriculture: selected references. Compiled . . . under the direction of Mary G[oodwin] Lacy. Department of agriculture: Bureau of agricultural economics: Agricultural economics bibliography (no.41): Washington 1933. pp.[ii].ii.51. [191.]*

LOUISE O[LDHAM] BERCAW, MARGARET T[HOMPSON] OLCOTT and MARY F[RANCES] CARPENTER, State measures for the relief of agriculture indebtedness in the United States, 1932 and 1933. Department of agriculture: Bureau of agricultural economics: Agricultural economic bibliography (no.45): Washington 1933. pp.[ii].68. [200.]*

— — 1933–1934. Compiled by M. T. Olcott and Louise O[ldham] Bercaw. . . . (no.53): 1934. pp.[iv].406. [250.]*

ESTHER M[ARIE] COLVIN, Part-time farming. A brief list of recent references. Department of agriculture: Bureau of agricultural economics: Agricultural economic bibliography (no.43):

Washington 1933. pp.[ii].23. [56.]*

— — [another edition]. Compiled by Helen E[mma] Hennefrud. . . . (no.77): 1939. pp.vi.272. [777.]*

M. A. DOWNES, Special administrative list of publications of the United States department of agriculture available free to agricultural administrators, investigators and instructors for their official use. Washington 1934. pp.ii.20. [500.]

MARY G[OODWIN] LACY, *ed.* The american farm problem. A selected list of books and pamphlets on the economic status of the farmer and measures for his relief since 1920. Department of agriculture: Bureau of agricultural economics: Agricultural economic bibliography (no.52): Washington 1934. pp.[ii].17. [100.]*

LIST of publications and other printed material of the Agricultural adjustment administration, May 12, 1933 to June 1, 1934. Department of agriculture: General information series (G–19): Washington 1934. pp.iii.25. [400.]

LOUISE O[LDHAM] BERCAW and HELEN E[MMA] HENNEFRUND, Farm tenancy in the United States, 1925–1935. Department of agriculture: Bureau of agricultural economics: Agricultural economics bibliography (no.59): Washington 1935. pp.[ii].ii.90. [298.]*

— — [another edition]. Farm tenancy . . . 1918–1936. . . . Compiled by L. O. Bercaw under the direction of Mary G. Lacy. . . . (no.70): 1937. pp.viii..302. [1070.]*

— — — Farm tenancy . . . 1937–1939. . . . Compiled by John M[ilton] McNeill under the direction of M. G. Lacy. . . . (no.85): 1940. pp.vi.160. [433.]*

— — — Farm tenancy . . . 1940–1941. Department of agriculture: Library list (no.2): Washington 1942. pp.49. [255.]*

ESTHER M[ARIE] COLVIN and JOSIAH C[HASE] FOLSOM, Agricultural labor in the United States, 1915–35. A selected list of references. Department of agriculture: Bureau of agricultnral economics: Agricultural economics bibliography (no.64): Washington 1935. pp.vii.493. [1746.]*

— 1936–1937. . . . (no.72): 1938. pp.[i].vii.205. [537.]*

— 1938–June 1941. Compiled by John M. McNeill and Josiah C. Folsom under the direction of Margaret T. Olcott. . . . (no.95): 1942. pp.iii–vi.268. [690.]*

— July 1941–February 1943. Compiled by Ruby W. Moats and J. M. McNeill. Department of agriculture: Library: Library list (no.4): 1943. pp.59. [289.]*

AGRICULTURAL relief measures relating to the raising of farm prices. 70th [&c.] congress. Department of agriculture: Bureau of agricultural economics: Library: Washington.*

70th. Compiled by Louise O. Bercaw. 1935. pp.27. [35.]

71st. Compiled by Vajen H. Fischer. 1935. pp.76. [90.]

72nd. 1935. pp.59. [68.]

73rd. Compiled by Marian E[tta] Wheeler and Mamie I[della] Herb. Department of agriculture: Bureau of agricultural economics: Agricultural economic bibliography (no.92): 1941. pp.iii–vi.80. [107.]

74th. Compiled by Marian E. Wheeler. . . . (no.84): 1940. pp.vi.75. [103.]*

75th. Compiled by M. E. Wheeler and M. I. Herb. . . . (no.76): 1939. pp.v.109. [138.]

ESTHER M[ARIE] COLVIN, Transportation of agricultural products in the United States, 1920–June 1939. A selected list of references to the various phases of railway, motor, and water carrier transportation. Department of agriculture: Bureau of agricultural economics: Agricultural economics bibliography (no.81): Washington 1939. pp.x. 250+viii.251+565+[iii].566–812. [4073.]

HAROLD O[RDELL] THOMEN, Checklist of hearings before congressional committees through the sixty-seventh congress. Part 1–House committee on agriculture. Preliminary edition. Library of Congress: Legislative reference service: Washington 1941 [1942]. pp.[ii].54. [500.]*

*a revised edition is entered below under United States: Congress.*

POPULAR publications for the farmer and homemaker. Department of agriculture: Office of information: Division of publications: List (no.5): [Washington 1944]. pp.31. [1350.]

*limited to the Department's publications.*

WALTER T[HEODORE] BORG, War and agriculture in the United States, 1914–1941. Selected references. Department of agriculture: Bureau of agricultural economics: Agricultural economics bibliography (no.93): Washington 1942. pp.iii–iv.43. [230.]*

SELECTED list of american agricultural books. [Second edition]. Department of agriculture: Library list (no.1): [Washington 1942]. pp.29. [400.]*

*revised in 1944.*

LAURA ISABEL MAKEPEACE, References on agriculture for returning veterans interested in farming. Colorado A & M college library: Library bulletin (no.13): Fort Collins 1945. ff.11. [190.]*

HAROLD T. PINKETT, Preliminary inventory of the records of the Office for agricultural war relations. National archives: [Preliminary inventories (no.37):] Washington 1952. pp.v.19. [large number.]*

HAROLD T. PINKETT, Preliminary inventory of the records of the Bureau of plant industry, soils and agricultural engineering. National archives: Preliminary inventories (no.66): Washington 1954. pp.v.49. [very large number.]*

AGRICULTURE in the United States. Books available . . . in the American library. 1954. pp.[v].34. [500.]*

SELECTED references on migratory agricultural workers and their families—problems and programs 1950 to April 1954. Department of labor: Bureau of labor standards: Washington 1954. pp.7. [75.]*

— 1950–1955. 1955. pp.[ii].11. [100.]*

VIRGIL E. BAUGH, Preliminary inventory of the records of the Extension service. National archives: Preliminary inventories (no.83): Washington 1955. pp.v.37. [large number.]*

AGRICULTURE and the allied sciences. United States information agency: Information center service: Subject bibliography (no.53): [Washington] 1959. pp.[ii].45. [700.]*

SELECTED list of american agricultural books in print and current agricultural periodicals. Department of agriculture: Library list (no.1): Washington 1960. pp.34. [1250.]

b. *Specific places &c.*

SAMUEL L. BOARDMAN, Agricultural bibliography of Maine. Augusta 1893. pp.117. [1250.]

EARL G. SWEM, A list of manuscripts relating to the history of agriculture in Virginia, collected by N[athaniel] F[rancis] Cabell, and now in the Virginia state library. Virginia state library: Bulletin (vol.vi, no.1): Richmond 1913. pp.20. [133.]

LIST of references on abandoned farms in New England. Library of Congress: Washington 1913. ff.6. [83.]*

N. F. CABELL, A contribution to the bibliography of agriculture in Virginia. Edited by Earl G.

Swem. Virginia state library: Bulletin (vol.xi, nos.1–2): Richmond 1918. pp.35. [2500.]

MARGARET T[HOMPSON] OLCOTT, Alabama. An index to the state official sources of agricultural statistics. Department of agriculture: Bureau of agricultural economics: Agricultural economic bibliography (no.15): Washington 1926. pp.[ii]. iii.98. [1750.]*

ICELLE E[MMA] WRIGHT, Oklahoma. An index to the state official sources of agricultural statistics. . . . Including a list of the unofficial sources of Oklahoma agricultural statistics, compiled by Margaret Walters. Department of agriculture: Bureau of agricultural economics: Agricultural economics bibliography (no.21): Washington 1927. pp.[ii].iv.462. [9000.]*

LOUISE O[LDHAM] BERCAW, California. An index to the state sources of agricultural statistics. Department of agriculture: Bureau of agricultural economics: Agricultural economics bibliography (no.31): Washington*.

i. Fruits, vegetables and nuts. 1930. pp.[ii]. xlii.724. [10,000.]
ii. Crops other than fruits, vegetables and nuts. 1930. pp.[ii].xlii.430. [6000.]

iii. Livestock and livestock products. 1931. pp.[ii].xlii.371. [5000.]

iv. Land, farm property, irrigation, and miscellaneous items. 1931. pp.[ii].xlii.128. [2000.]

v. An index to some unofficial sources. Compiled under the direction of M[ary] J[essamine] Abbott. 1930. pp.[ii].x.69. [1250.]

LIST of publications [Available bulletins]. Agricultural [and home economics] experiment station: Ames, Iowa 1936 &c.

*in progress.*

HELEN E[MMA] HENNEFRUND and ORPHA CUMMINGS, Bibliography on the Japanese in american agriculture. Department of agriculture: Bibliographical bulletin (no.3): Washington 1943 [1944]. pp.iv.61. [393.]

[D.] MATTHEWS, An author and subject index to the publications of the Georgia agricultural experiment station and the Georgia coastal plain experiment station, 1888–1946. University of Georgia: Libraries: [Athens] 1948. ff.[ii].138. [2500.]*

— — Supplement. Compiled by Virginia Michaelis. 1956. ff.77. [1500.]

## *Agriculture*

RICHARD BARDOLPH, Agricultural literature and early Illinois farmers. Illinois studies in the social sciences (vol. xxix, nos.1–2): Urbana 1948. pp.200. [400.]

ORPHA CUMMINGS, Economic development of the western range resources. An inventory of research. Montana state college: Bozeman 1952. pp.[iv].60. [600.]*

LIST of publications and patents of the Southern utilization research branch. Department of agriculture: Agricultural research administration: Southern utilization research branch: New Orleans 1954 &c.*

*in progress.*

LIST of publications and patents with abstracts. Agricultural research service: Western utilization research branch: Albany, Cal. 1962 &c.*

*in progress.*

### *Venezuela*

PEDRO GRASES, Contribución a la bibliografía venezolana de temas agropecuarios. Ministerio de agricultura y cría: Publicaciones: Caracas 1943. pp.3–180. [469.]

V[ICTOR] M. BADILLO and C[ELESTINO] BONFANTI, Indice bibliográfico agrícola de Venezuela. Fundación Eugenio Mendoza: Caracas 1957. pp.xviii. 305. [5822.]

*Wales*

[D. LLEUFER THOMAS], Bibliographical, statistical, and other miscellaneous memoranda, being appendices to the report of the Royal commission on land in Wales. 1896. pp.[iii].205. [4000.]

*Yugoslavia*

SAVREMENA jugoslovenska poljoprivredna bibliografija za 1945 g. Ministarstvo poljoprivrede i šumarstva FNJR: Beograd 1947. pp.xl.63. [718.]

[*continued as:*]

Poljoprivredna bibliografija FNRJ.

1946. pp.124.xii. [1508.]
1947. pp.134. [1524.]
1948. pp.160. [1496.]
1949. pp.195. [1632.]
— Dodatak za 1945, 1946, 1947, 1948, 1949. 1950. pp.74.

КАТАЛОГ пољопривредних књига. Београд 1950. pp.109. [500.]

## 7. *Agricultural appliances and engineering*

ABRIDGMENTS of specifications. Class 5. Agricultural appliances, farmyard and like. Patent office.

1855–1866. pp.xii.76. [300.]
1867–1876. pp.xii.82. [400.]
1877–1883. pp.x.68. [300.]
1884–1888. pp.x.86. [300.]
1889–1892. pp.xii.80. [300.]
1893–1896. pp.xii.78. [300.]
1897–1900. pp.xiv.98. [300.]
1901–1904. pp.xiv.108. [400.]
1905–1908. pp.xiv.130. [500.]

ABRIDGMENTS of specifications. Class 6, agricultural appliances for the treatment of land and crops. Patent office.

1855–1866. pp.xviii[*sic,* xxviii].342. [1250.]
1867–1876. pp.xxv.426. [1500.]
1877–1883. pp.xviii.246. [750.]
1884–1888. pp.xvi.200. [600.]
1889–1892. pp.xvi.160. [500.]
1893–1896. pp.xvii.144. [500.]
1897–1900. pp.xx.146. [500.]
1901–1904. pp.xxii.224. [750.]
1905–1908. pp.xxiii.250. [1000.]

ABRIDGMENTS of specifications relating to agriculture. Commissioners of patents for specifications.

i. Field implements (including methods of tilling and irrigating land), A.D. 1618–1866. 1876. pp.xix.931. [1000.]
— 1867–1876. 1878. pp.xvi.492. [500.]
ii. Barn and farmyard implements (including the cleaning, drying, and storing of grain), A.D. 1636–1866. 1879. pp.xv.480. [750.]
— 1867–1876. 1879. pp.xii.252. [500.]
iii. Agricultural and traction engines, A.D. 1618–1866. 1884. pp.viii.199. [250.]
*no more published.*

SELECT list of references on agricultural implements, trade and manufacture. Library of Congress: Division of bibliography: [Washington] 1911. pp.9. [70.]*

V[ASILY PROKHOROVICH] GORYACHKIN, Опытъ перечня книгъ по земледѣльческой механикѣ. Бюро по сельско-хозяйственной механикѣ: С.-Петербургъ 1913. pp.[ii].iii.96. [1685.]

PATENTS for inventions. Fifty years subject index, 1861–1910. Class 6 (i). Cultivating implements and systems. Patent office: 1915. pp.19. [2500.]

PATENTS for inventions. Fifty years subject

index, 1861–1910. Class 6 (iii). Harvesting-appliances. Patent office: 1914. pp.15. [2000.]

ABRIDGMENTS of specifications. Class 5 (i). Farmyard and like appliances (other than housing and feeding animals). Patent office.

1909–1915. pp.viii.96. [300.]
1916–1920. pp.vii.36. [100.]
1921–1925. pp.viii.72. [200.]
1926–1930. pp.viii.77. [250.]
*no more published.*

ABRIDGMENTS of specifications. Class 6 (i). Cultivating implements and systems. Patent office.

1909–1915. pp.xiii.216. [600.]
1916–1920. pp.xiv.200. [600.]
1921–1925. pp.xiv.169. [600.]
1926–1930. pp.xiii.141. [500.]
*no more published.*

ABRIDGMENTS of specifications. Class 6 (iii). Harvesting-appliances. Patent office.

1909–1915. pp.ix.151. [500.]
1916–1920. pp.viii.76. [250.]
1921–1925. pp.viii.100. [250.]
1926–1930. pp.viii.92. [250.]
*no more published.*

## *Agriculture*

LIST of references on agricultural machinery in India. Library of congress: Division of bibliography: [Washington] 1922. pp.6. [75.]*

ABSTRACTS of current literature and notes. Institute for research in agricultural engineering: Oxford.

i–ii. 1931. ff.12.14. [100.]
iii–vi. 1932. ff.14.15.13.16. [250.]
vii–ix. 1933. ff.14.14.11. [200.]
x–xiii. 1934. ff.13.15.14.15. [250.]
— [index]. Nos.1–13. ff.27.
xiv–xvii. 1935. ff.12.10.15.16. [250.]
xviii–xxi. 1936. ff.21.21.10.16. [350.]
xxii–xxiv. 1937. ff.14.18.15. [200.]

*no more published.*

AGRICULTURAL engineering current literature. Department of agriculture: Bureau of agricultural [chemistry and] engineering: Washington.*

i. 1931–1932. pp.31.32.22.21.19.ff.32. pp.29.17.23.30.15. [2500.]
ii. 1932–1933. pp.30.21.15.18.20.17.14.15.13.19.15.20.17.22.20. [2500.]
iii. 1933–1934. pp.27.29.22.20.19.21.20.24.19.18.24.24. [2500.]
iv. 1934–1935. pp.17.21.21.20 [*sic*, 21].19.18.21.24.23.18.24.16. [2500.]

v. 1935–1936. pp.17.24.22.24.20.23.18.20.21. 21.24.24. [2500.]

vi. 1936–1937. pp.20.21.24.24.24.24.21.24.23. 21.24.23. [2500.]

vii. 1937–1938. pp.21.24.21.24.24.24.24.22. 23.24.24.24. [2500.]

viii. 1938–1939. pp.22.24.24.21.24.24.20.21. 24.24.24.24. [2500.]

ix. 1939–1940. pp.24.24.24.24.24.21.18.21.20. 20.22.24. [2500.]

x. 1940–1941. pp.24.18.20.19.17.20.24.21.23. 20.24. [2500.]

*no more published.*

AGRICULTURAL wheels, 1930–5. Science library: Bibliographical series (no.228): 1936. ff.3. [53.]*

DOROTHY W[ILLIAMS] GRAF, Agricultural engineering, a selected bibliography. Department of agriculture. Bureau of agricultural engineering: Washington 1937. pp.373. [1550.]*

SELECT list of references on harvesting machinery, 1931 onwards. Science library: Bibliographical series (no.562): 1941. ff.2. [40.]*

RICHARD WIEBE and JANINA NOWAKOWSKA, The technical literature of agricultural motor fuels. Department of agriculture: Bibliographical bulletin (no.10): Washington 1949. pp.vi.259. [1562.]

GUSTAV BERNSTRÖM, Jacob Pontus von Wulfschmidts Bondestolpe, en bibliografisk granskning. Uppsala 1950. pp.121. [15.]

AUSZÜGE deutscher patent-anmeldungen. Landwirtschaft. Klasse 45,50,66,64. [Berlin 1947–1948]. pp.641. [4487.]

*covers the period 1939–1945; issued by the allied occupation forces.*

ABSTRACTS on agricultural and horticultural engineering. National institute of agricultural engineering: Silsoe.

[i]. 1950–1951. [654.]

[*continued as:*]

Agricultural and horticultural engineering abstracts.

ii. 1951–1952. pp.75.85.68.81.29. [599.]

iii. 1952–1953. pp.87.71.76.80.35. [937.]

*in progress.*

NATHAN REINGOLD, Preliminary inventory of the records of the Bureau of agricultural engineering. National archives: Preliminary inventories (no.53): Washington 1953. pp.v.28. [large number.]*

FÖLDMUNKÁK gépesítése. Országos műszaki könyvtar: Iparági bibliográfiái (no.14): Budapest 1953. pp.68. [491.]

L. M. VADIKOVSKAYA [*and others*], MTC — решающая колхозного производства. Рекомендательный указатель литературы по механизации сельского хозяйства в помощь работникам МТС. Государственная . . . публичная библиотека СССР имени В. И. Ленина: Библиографический отдел: Москва 1954. pp.82. [350.]

LIST of publications. National institute of agricultural engineering and Scottish machinery testing station: Silsoe &c. 1955. pp.13. [400.]

A. A. KHRENKOVA and N. M. METELKIN, Работа библиотек в помощь механизаторам сельского хозяйства. Сборник материалов. Государственная . . . библиотека СССР имени В. И. Ленина: Научно-методический кабинет библиотековедения: Москва 1957. pp.164. [250.]

MECHANIZACE a elektryfikace zemědělství. (Soupis literatury). Státní techniká knihovna: Brno 1959. pp.[ii].57. [400.]*

G. N. SINEOKOV, *ed.* Библиографический справочник по почвообрабатывающим машинам и орудиям. Всесоюзный научно-исследовательский институт сельскохозяйственного машиностроения: Москва.

i.

ii. Литература на русском языке за 1930–1955 гг. 1959. pp.264. [2500.]

*in progress.*

MECHANIZÁCIA v poľnohospodárstve. Zoznam odporúčanej literatúry. Ústredná techniská knižnica: Bratislava 1959. pp.31. [75.]

S. E. KAPLAN and A[LEKSANDR] E[VGENEVICH] CHZHAO, Механизация и электрификация сельского хозяйства. Рекомендательный указатель литературы. Государственная ... библиотека СССР имени В. И. Ленина and Центральная сельскохозяйственная библиотека ВАСХНИЛ: Москва 1960. pp.115. [600.]

ELIZABETH GOULD DAVIS, The economics of farm mechanization in the United States, 1950–1960. Department of agriculture: Library list (no.68): Washington 1960. pp.16. [262.]*

CARL W. HALL, Agricultural engineering index, 1907-1960. Agricultural consulting associates [Reynoldsburg, O.]: Ann Arbor 1961. pp.xix.135. [9000.]*

## 8. *Agricultural chemistry*

JAHRESBERICHT über die fortschritte der agriculturchemie, mit besonderer berücksichtigung der pflanzenchemie und pflanzenphysiologie. Berlin.

i. 1858–1859. pp.v.249. [500.]
ii. 1859–1860. pp.iv.330. [500.]
iii. 1860–1861. pp.iv.292. [500.]
iv. 1861–1862. pp.[ii].300. [500.]
v. 1862–1863. pp.[ii].268. [500.]
vi. 1863–1864. pp.[ii].235. [500.]

[*continued as:*]

Jahresbericht über die fortschritte auf dem gesammtgebiete der agrikultur-chemie.

vii. 1864. Edited by Eduard Peters. pp.iv.445. [1000.]
viii. 1865. pp.444. [1000.]
ix. 1866. pp.[ii].510. [1000.]
x. 1867. pp.[ii].404. [1000.]
xi–xii. 1868–1869. Edited by Th. Dietrich [*and others*]. pp.xxviii.776. [2000.]
xiii–xv. 1870–1872. pp.ix.284+x.240+xix.334. [3000.]
xvi–xvii. 1873–1874. pp.xv.352+xv.308+vii.124. [2000.]
xviii–xix. 1875–1876. pp.xxvi.482+xix.312+vii.92. [3000.]

xx. 1877. pp.xxxiv.730. [1500.]

— Generalregister über jahrgang 1–xx. Edited by A[lbert] Hilger. 1879. pp.viii.388.

*new series.*

i. 1878. Edited by A. Hilger. pp.xxxv.829. [2500.]

ii. 1879. pp.xxxii.685. [2000.]

iii. 1880. pp.xxxiv.680. [2000.]

iv. 1881. pp.xxix.624. [2000.]

v. 1882. pp.xxxi.612. [2000.]

vi. 1883. pp.xxix.614. [2000.]

vii. 1884. pp.xxix.758. [2000.]

viii. 1885. pp.xxxii.666. [2000.]

ix. 1886. pp.xxxiv.638. [2000.]

x. 1887. pp.xxx.680. [2000.]

xi. 1888. pp.xxx.606. [2000.]

xii. 1889. pp.xxx.710. [2000.]

xiii. 1890. Edited by A. Hilger and Th. Dietrich. pp.xxxvii.850 [2500.]

xiv. 1891. pp.xxxviii.756. [2500.]

xv. 1892. pp.xxxviii.758. [2500.]

xvi. 1893. pp.xxxii.556. [1000.]

xvii. 1894. pp.xxxvi.698. [2000.]

xviii.1895. pp.xxxv.668. [2000.]

xix. 1896. pp.xxviii.712. [2000.]

xx. 1897. pp.xl.766. [2000.]

— Generalregister zu band 1–xx der neuen

folge. By Heinrich Kraut and Th. Dietrich. 1899–1908. pp.iv.278+279–510+[iv].548.

*third series.*

i. 1898. pp.xxxviii.700. [2000.]
ii. 1899. pp.xxxvi.666. [2000.]
iii. 1900. pp.xxxvi.718. [2000.]
iv. 1901. pp.xxxviii.614. [2000.]
v. 1902. pp.xxxvi.580. [2000.]
vi. 1903. pp.xxxvi.662. [2000.]
vii. 1904. Edited by Th. Dietrich, pp.xliv. 740. [2000.]
viii. 1905. pp.xxxvi.562. [2000.]
ix. 1906. pp.xxxviii.626. [2000.]
x. 1907. pp.xxxv.624. [2000.]
xi. 1908. pp.xxxvi. 656. [2000.]
xii. 1909. pp.xxxiv.546. [2000.]
xiii. 1910. pp.xxxii.620. [2500.]
xiv. 1911. pp.xxxiv.676. [3000.]
xv. 1912. pp.xxx.556. [2000.]
xvi. 1913. pp.xxxi.564. [2000.]
xvii. 1914. Edited by Th. Dietrich and F. Mach. pp.xxix.564. [2000.]
xviii. 1915. pp.xxxii.504. [2000.]
xix. 1916. Edited by F. Mach. pp.xxxvi.520. [2000.]
xx.1917. pp.xxx.514. [2000.]

[*continued as:*]

Jahresbericht für agrikultur-chemie.
*fourth series.*

i. 1918. pp.xxx.546. [2000.]
ii. 1919. pp.xxx.564. [2000.]
iii. 1920. pp.xxxii.584. [2500.]
iv. 1921. pp.xxxi.578. [2500.]
v. 1922. pp.xxvii.524. [2500.]
vi. 1923. pp.xxix.564. [3000.]
vii. 1924. pp.xxix.584. [3000.]
viii. 1925. pp.xxx.576. [3000.]
ix. 1926. pp.xxxiii.640. [3500.]
x. 1927. pp.xxxii.574. [3500.]
xi. 1928. pp.xxxv.572. [3500.]
xii. 1929. pp.xxxiv.586. [3500.]
xiii. 1930. pp.xxxix.608. [4000.]
xiv. 1931. pp.xxxvi.566. [4000.]
xv. 1932. pp.[ii].ii.532. [4500.]
xvi. 1933. pp.iv.539. [5000.]

F. HONCAMP and O[TTO] NOLTE, Agrikulturchemie. Wissenschaftliche forschungsberichte: Naturwissenschaftliche reihe (vol.x): Dresden &c. 1924. pp.viii.160. [1250.]

LITERATURSAMMLUNG aus dem gesamtgebiet der agrikulturchemie. München [Weihenstephan; Leipzig].

i. Bodenkunde. Von H[ans] Niklas, F.

Czibulka und A[lbert] Hock. 1931.
ii. Bodenuntersuchung.
iii. Pflanzenernährung. Von H. Niklas [*and others*].
iv. Düngung und düngemittel.
v. Ergänzungsband zu band IV. Von H. Niklas und F. Ader. [1939]. pp.xxvii.472. [8500.]

H[ERBERT] P[AUL] HOLMAN [*and others*], Index of publications of the Bureau of chemistry and soils, originally the Bureau of chemistry and the Bureau of soils . . . 1862–1937. Volume I. List of titles and authors. Department of agriculture: Bureau of chemistry and soils: [Washington] 1937 [1939]. pp.[vii]. 546. [8428.]*
*no more published.*

LIST of processed publications of the Bureau of agricultural and industrial chemistry. [Washington] 1953. pp.[iii].25. [355.]*

I[VAN] I[LLARIONOVICH] RESHETINSKY, Химия в борьбе за урожай. Рекомендательный обзор литературы. Государственная... библиотека СССР имени В. И. Ленина: Центральная политехническая библиотека: Новости науки и техники (no.25): Москва 1960. pp.24. [27.]

INDUCTION de mutations chez les plantes supérieures d'intérêt agricole par les rayonnements bêta, gamma, les neutrons et les protons, 1955–1960. Commissariat à l'énergie atomique: Isotopes, rayonnements, agriculture: Études bibliographiques (no.1): Gif-sur-Yvette 1962. pp.xv.282. [439.]

9. *Agricultural economics and productivity*

AGRICULTURAL credit. [List of United States documents on agricultural credit banks]. Bibliography. Library of Congress [Washington] 1915. ff.10. [88.]*

LIST of United States documents on agricultural credit banks. Library of Congress: Division of bibliography: [Washington] 1916. pp.9. [80.]*

FARM management. Publications relating to farm accounts, credits, marketing, homes, and statistics for sale. Superintendent of documents: Price list (no.68): Washington 1916. pp.24. [500.]
— 38th edition. Farm management, foreign agriculture, rural electrification. 1953. pp.14. [350.]

BRIEF list of references on the outlook for the farmer (methods of marketing, increasing productivity of the soil, financing, labor supply and substitution of machinery, etc.). Library of Congress: [Washington] 1921. ff.10. [130.]*

LIST of recent references on the factors influencing the prices of staple agricultural products. Library of Congress: Washington 1921. ff.11. [127.]*

CHASTINA GARDNER, Agricultural cooperation. A selected and annotated reading list, with special reference to purchasing, marketing, and credit. Including only works printed in english, and exclusive of periodical references except reprints and proceedings of associations. Department of agriculture: Miscellaneous circular (no.11): Washington 1923. pp.55. [350.]

— — [Fourth edition]. Farm credit administration: Cooperative division: Bulletin (no.4): Washington 1936. pp.iii.214. [1050.]

A LIST of references on agricultural credit. Library of Congress: [Washington] 1924. ff.14. [162.]*

EMILY L. DAY, KATHARINE JACOBS and MARGARET T. OLCOTT, Bibliography on the marketing of agricultural products. Department of agriculture: Bibliographical contributions (no.7): Washington 1924. ff.v. pp.133. [1750.]*

— — [another edition]. Department of agriculture: Miscellaneous circular (no.35): 1925. pp.[iv].56. [1750.]

— — — [supplement]. Compiled by Louise O[ldham] Bercaw and Esther M[arie] Colvin . . . under the direction of Mary G[oodwin] Lacy. Department of agriculture: Miscellaneous Publication (no.150): 1932. pp.iv.351. [3039.]

MARY G[OODWIN] LACY, Agricultural economics. A selected list of references. Department of agriculture: Bureau of agricultural economics: Agricultural economics bibliography (no.1): Washington 1925. ff.[i].11. [150.]*

— — [another edition]. 1938. pp.lv.26. [300.]

LOUISE O[LDHAM] BERCAW, Price spreads. A selected list of references relating to analyses of the portion of the consumer's price accruing to various agencies. Department of agriculture: Bureau of agricultural economics: Agricultural

LIST of speeches, &c., on federal agricultural relief as printed in the Congressional Record, 69th Congress, first session, v.67, current file. Library of Congress: Washington 1926. ff.17. [279.]*

LIST of recent references on federal agricultural

economics bibliography (no.4): Washington 1925. ff.[i].20. [200.]*

relief. Library of Congress: Washington 1926. ff.13. [172.]*

A[NNIE] M[URRAY] HANNAY, Bounties on agricultural products. A selected bibliography. Department of agriculture: Bureau of agricultural economics: Agricultural economic bibliography (no.20): Washington 1927. pp.[iv].128. [343.]*

B. A. E. news. Library supplement. Department of agriculture: Bureau of agricultural economics: Washington.

[*continued as:*]

Agriculture economics literature.*

i. 1927. pp.323. [1000.]
ii. 1928. pp.417. [1000.]
iii. 1929. pp.556. [2500.]
iv. 1930. pp.740. [3500.]
v. 1931. pp.838. [4000.]
vi. 1932. pp.884. [4500.]
vii. 1933. pp.954. [5000.]
viii. 1934. pp.1033. [5000.]
ix. 1935. pp.1229. [6000.]
x. 1936. pp.1229. [6000.]
xi. 1937. pp.1437. [7000.]
xii. 1938. pp.1563. [7500.]

xiii. 1939. pp.1615. [7500.]
xiv. 1940. pp.1588. [7500.]
xv. 1941. pp.1218. [6000.]
xvi. January-June 1942. pp.536. [3000.]
*merged into the* Bibliography of agriculture.

MARGARET T[HOMPSON] OLCOTT, Taxation and the farmer. A selected and annotated bibliography. Department of agriculture: Bureau of agricultural economics: Agricultural economics bibliography (no.25): Washington 1928. pp.[ii]. v.190. [742.]*

LOUISE O[LDHAM] BERCAW, Agricultural relief. A selected list of references. Department of agriculture: Bureau of agricultural economics: Washington 1928. pp.23. [200.]*

— — [second edition]. Agricultural relief. A selected and annotated bibliography. Compiled by Margaret T. Olcott. Department of agriculture: Bureau of agricultural economics: Agricultural economics bibliography (no.50): Washington 1933. pp.iv.382. [2252.]*

ANTON BROSCH, Deutsches schrifttum der agrarpolitik und weltlandwirtschaft von 1920–1929. Reichsministerium für ernährung und landwirtschaft: Berichte über landwirtschaft (new ser., special no.26): Berlin 1930. pp.xii. coll.350. pp.351–374. [7500.]

ESTHER M[ARIE] COLVIN, List of state official serial publications containing material on agricultural economics. Department of agriculture: Bureau of agricultural economics: Agricultural economics bibliography (no.38): Washington 1932. pp.[ii]. 222 [*sic*, 223]. [205.]*

A[NNE] M[ARIE] H[ANNAY], MARY COULT and LILLIAN CRANS, Measures taken in foreign countries to relieve agricultural indebtedness. Department of agriculture: Bureau of agricultural economics: Agricultural economic bibliography (no.42): Washington 1933. pp.[iii].ii.57. [127.]*

VAJEN E[ILLEEN] HITZ, Business and agriculture, 1920–1933. A parial bibliography of material on the interdependence of business and agriculture. Department of agriculture: Bureau of agricultural economics: Agricultural economic bibliography (no.51): Washington 1933. pp.[ii].iv.155. [448.]*

CHRONOLOGICAL list of department publications. University of Reading: Agricultural economics department: [Reading].*

1924–1937. 1937. ff.[i].5. [50.]
1937–1947. 1948. ff.[i].5. [50.]

ORPHA CUMMINGS, Rural land economics, 1936

... relating ... to the present national land policy. [Berkeley] 1937. ff.93. [1170.]*

PRINTED publications issued by the Bureau of agricultural economics. Department of agriculture: Bureau of agricultural economics: Washington 1937. pp.[ii].23. [360.]*

AGRICULTURAL relief measures relating to the raising of farm prices. Department of agriculture: Bureau of agricultural economics: Agricultural economics bibliography (no.92 &c.): Washington.

1933–1934. Compiled by Marion E. Wheeler and Marie I. Herb ... (no.92): 1941. pp. iii.–vi.80. [107.]*
1935–1936. ... (no.84): 1940. pp.vi.75. [103.]
1937–1938. ... (no.75): 1939. pp.v.109. [140.]

BIBLIOGRAPHIE internationale d'économie rurale [&c.] Institut international d'agriculture [vol.viii, no.3 &c.: Food and agriculture organization of the United nations: European regional office]: Rome.

i. 1938–1939. pp.[xii].510. [7364.]
ii. 1939–1940, pp.430. [5849.]
iii. 1941. pp.367. [4990.]
iv. 1942. pp.351. [4649.]
v. 1943. pp.244. [3022.]
vi. 1944. pp.56. [467.]

vii. 1945. pp.51. [503.]
viii. 1946. pp.210. [2226.]

LOUISE O[LDHAM] BERCAW and HELEN E[MMA] HENNEFRUND, Distribution of farm income by size. A selected bibliography. Compiled . . . under the direction of Margaret T. Olcott. Department of agriculture: Bureau of agricultural economics: Agricultural economics bibliography (no.96): Washington 1942. pp.iii–iv.103. [269.]*

JOSEPH W. EATON and SAUL M. KATZ, Research guide on cooperative group farming. A research bibliography on rural cooperative production and cooperative communities. New York 1942. pp.86. [500.]

IMPORTANT sources of information for work in agricultural economics, with special emphasis on California. University of California: College of agriculture: Agricultural experiment station, Berkeley: [Berkeley] 1943. ff.[iii].77. [1200.]*
— Ninth edition. 1956. ff.[ii].108. [1500.]*

JOHN M. MCNEILL, Rehabilitation of low-income farmers. A list of references. Department of agriculture: Library: Library list (no.6): Washington 1943. pp.56. [258.]*

JOHN M. MCNEILL, Agricultural credit. Publications of the federal government and state agricultural colleges, 1929–1943. Department of agriculture: Library: Library list (no.7): Washington 1944. pp.41. [439.]*

HANDLIST of publications, 1924–1945. University college of Wales: Department of agricultural economics: Aberystwyth 1945. pp.[iv].33. [350.]*

AGRICULTURAL economics, with special emphasis on land: a brief list of recent references. Library of Congress: Washington 1946. ff.8. [40.]*

AGRICULTURAL, economic and statistical publications. Department of agriculture: Bureau of agricultural economics: Washington 1952. pp.ii.83. [2250.]*

LABOUR productivity in agriculture. International labour office: Library: Bibliographical reference list (no.57): Geneva 1952. pp.7. [100.]*

CHECKLIST of BAE publications. Department of agriculture: Bureau of agricultural economics: Division of economic information: Washington 1952 &c.*

*in progress.*

A LIST of published material by members of the Economics division, 1930–1953. Department of agriculture: Marketing service: Economics division: Ottawa 1953. pp.[v].44. [600.]*

*an earlier edition appeared in the* Economic annalist *(February 1950).*

— [another edition]. [By Helen I. Marquis]. List . . . 1930–1956. 1957. pp.[iii].64. [750.]*

CHECKLIST of reports and charts issued by the Agricultural marketing service. Department of agriculture: Agricultural marketing service: Washington 1953 &c.*

*in progress.*

PUBLICATIONS. Department of agriculture, Farmer cooperative service: Washington 1954. pp.29. [200.]*

ELIZABETH GOULD DAVIS, Low-income farm people. A selected list of references. Department of agriculture: Library list (no.62): Washington 1955. pp.[ii].46. [782.]*

GUNTHER HELBIG, Leistungssteigerung in der landwirtschaft. Ein wegweiser durch die fachliteratur. Empfehlende bibliographie. Deutsche bücherei: Sonderbibliographien (no.4): Leipzig [1956]. pp.69. [250.]

R. TRAUTMANN, Socialistická organisace práce v zemědělství. Universita: Knihovna: Výběrový seznam (no.10): [Brno] 1955. ff.[56]. [53.]

A. G. SUSHKO, Создадим обилие сельскохозяйственных продуктов. Беседа о книгах. Государственная... библиотека СССР имени В. И. Ленина: За новые успехи в шестой пятилетке: Москва 1956. pp.20. [13.]

SEMIANNUAL list [List] of publications and patents with abstracts. Department of agriculture: Agricultural research administration: Western utilization research branch: [Albany 1956] &c.*

*in progress.*

LIST of published material... by members of the Economic division. Department of agriculture: Ottawa.*

1930–1956. 1957. pp.[iii].64. [800.]

1957–1962. [By] Helen I. Marquis. 1964. pp. [v].40. [500.]

DIGEST of agricultural economics. University of Oxford: Institute for research in agricultural economics: [Oxford].

[1]–2. 1957. pp.39+50. [70.]

3–4. 1958. pp.56+56. [125.]

5–9. 1959. pp.51+90+87+46+39. [350.]

10–13. 1960. pp.43+53+60+39. [250.]

14–17. 1961. pp.42+32+50+47. [250.]
18–21. 1962. pp.44+56+58+48. [250.]
22–25. 1963. pp.36+64+53+56. [250.]

*in progress.*

VIVIAN WISER, Preliminary inventory of the records of the Bureau of agricultural economics. National archives: Preliminary inventories (no.104): Washington 1958. pp.vi.212. [very large number.]*

HANS JURGEN METZDORF, Veröffentlichungen auf dem gebiet der landwirtschaftlichen marktforschung in Westdeutschland. Land- und forstwirtschaftlicher forschungsrat: Bonn [1958]. pp.258. [3400.]

SOCIALIZÁCIA dediny v Prešovskom kraji. Bibliografia o socializácii dediny v Prešovskom kraji v rokoch 1948–1958. Státna vedecká knižnica: Prešov 1958. ff.[ii].66. [450.]*

ANNE DEWEES TAYLOR, A bibliographic guide to the writings of Henry C. Taylor, agricultural economist, covering the years 1893–1957. Agricultural history (vol.32, no.3, supplement): Champaign, Ill. 1958. pp.28. [308.]*

E. V. BAZHANOVA, Проблемы производительности труда и себестоимость продукции в сельском хозяйстве СССР после Великой отечественной войны. Библиографический указатель книг и журнальных статей за 1945–1957 гг. Академия наук СССР: Фундаментальная библиотека общественных наук: Ленинград &c. 1959. pp.160. [1274.]

STANLEY W. BROWN and VIRGIL E. BAUGH, Preliminary inventory of the records of the Commodity exchange authority. National archives: Preliminary inventories (no.112): Washington 1959. pp.v.33. [large number.]*

ELENA NIKITICHNA FOMINA [*and others*], За обилие продуктов сельского хозяйства. Обзоры литературы. Государственная . . . библиотека СССР имени В. И. Ленина: Семилетний план — план развернутого строительства коммунизма: Москва 1959. pp.71. [200.]

VIVIAN WISER, Writing relevant to farm management in the records of the Bureau of agricultural economics. National archives: Special lists (no.17): Washington 1963. pp.v.80. [large number.]*

### 10. *Agricultural education and research*

[EDWARD D. GREENMAN and EDITH A. WRIGHT], Bibliography of education in agriculture and home education. Bureau of education: Bulletin (no.481): Washington 1912. pp.62. [agriculture: 378.]

HARRY W[OODBURY] SMITH, *ed.* Annotated list of books recommended to secondary schools for use in teaching agriculture. University of Maine: College of agriculture: The Maine bulletin (vol. xvi, no.3): Waterville 1913. pp.24. [190.]

BRIEF list of references on agricultural education on Denmark, England, France, Germany. Library of Congress: Division of bibliography: [Washington] 1915. pp.3. [32.]*

CARL COLVIN, A bibliography of agricultural books for the high school library. Illinois: Board of vocational education: Bulletin (no.6): [Danville, Illinois 1919.] pp.31. [207.]

SUMMARIES of studies in agricultural education An annotated bibliography. American vocational association: Agricultural education section (= Office of education: Vocational education bulletin (no.180): Agricultural series (no.18 [*sic* for 47]): Washington 1935. pp.v.196. [373.]

— [another edition]. Federal security agency: Vocational division: Vocational division bulletin (no.242=Agricultural series, no.59): Washington 1950. pp.iv.61. [145.]

— — [supplement]. Office of education: Vocational division: Bulletin (no.265 = Agricultural series, no.69): pp.v.94. [186.]

J[OY] RICHARD BLANCHARD and HARALD OSTVOLD, Literature of agricultural research. Berkeley 1958. pp.x.231. [1500.]*

*Agricultural Entomology,*
see **Entomology: Agricultural**

11. *Agricultural meteorology*

BIBLIOGRAPHY of literature on agricultural meteorology. Ministry of agriculture and fisheries and Board of agriculture for Scotland: Agricultural meteorological scheme.*
- i. 1926. ff.[21]. [3000.]
- ii. 1927. pp.[iv].33. [450.]

AGRARMETEOROLOGISCHE bibliographie.
- 1949. Von Max Schneider. Deutscher wetterdienst in der US-zone: Bad Kissingen 1950–1952. pp.vi.229.xviii.255. [2235.]*
- 1950. Bibliographien des Deutschen wetterdienstes (no.1): 1955. pp.xii.129. [1543.]
- 1951. . . . (no.3): 1955. pp.x.96. [861.]
- 1952. . . . (no.4): 1956. pp.xii.107. [946.]
- 1953. . . . (no.5): 1956. pp.x.86. [846.]

1954. . . . (no.6): 1957. pp.x.82. [777.]
1955. . . . (no.7): 1957. pp.ix.64. [520.]
1956. . . . (no.8): Offenbach/M. 1958. pp.x. 119. [972.]
1957. . . . (no.9): 1958. pp.x.92. [662.]
1958. . . . (no.10): 1960. pp.xiii.235. [1485.]
1959. . . . (no.11): 1961. pp.xv.265. [1830.]
1960. . . . (no.12): 1962. pp.xvi.265. [1944.]
1961. . . . (no.13): 1963. pp.xiii.253. [1471.]

*in progress.*

GERALD L. BARGER, A selected bibliography of agricultural meteorology. Weather bureau: Washington 1956. ff.[ii].99. [1500.]*

J. L. DIONNE, GEO[RGE] W. ROBERTSON and R. M. HOLMES, Agricultural meteorology. A brief literature review & bibliography. Canada department of agriculture: Experimental farms service: [Ottawa] 1956. pp.[iv].59. [578.]*

ELIZABETH GOULD DAVIS, A selected bibliography on weather-crop relations in the great plains. Department of commerce: Weather bureau [&c.]: [Washington] 1956. pp.16. [200.]*

BIBLIOGRAFIA z rakresu meteorologii rolniczej i leśnej. Państwowy instytut hydrologiczmo-meteorologiczny: Warszawa 1958. pp.124. [600.]

JEN-YU WANG, *ed.* Bibliography of agricultural meteorology at the university of Wisconsin. University of Wisconsin: Department of meteorology and soils: Madison 1960. ff.[ii].v.41. [500.]*

ANNE ZIMMERMANN and JEN-YU WANG, Bibliography for agrometeorological studies on south America. University of Wisconsin: Department of meteorology and soils: [Madison 1960]. ff. iii.23. [250.]*

## 12. *Agricultural statistics*

GEO. F. THOMPSON and GEO. W. M. HILL, Synoptical index of the reports of the statistician, 1863 to 1894. Department of agriculture: Division of publications: Bulletin (no.2): Washington 1897. pp.258. [5000.]

KORNÉL TURÁNYI, A mezőgazdasági statisztika legújabb forrásai. Az agrártudományi egyetem központi könyvtárának kiadványai (4th ser., no.4) Budapest 1962 &c.

*in progress.*

## 13. *Miscellaneous*

ABRIDGMENTS of the specifications relating to steam culture. Patent office: 1858. pp.vi.130. [150.]

*covers the period 1618–1856.*

[EUGÈNE GAYOT], Travaux agronomiques de m. Eug. Gayot. [1863]. pp.3. [18.]

[ALEXANDRE LABOULBÈNE], Notice sur les principaux travaux scientifiques du dr A. Laboulbène se raportant aux questions relatives à l'économie rurale. [1894]. pp.75. [235.]

DRY farming. Public library: Bulletin (no.40): Riverside, Cal. 1912. pp.12. [125.]

SELECT list of references on conservation on the farm: prevention and utilisation of waste, application of power, scientific management, etc. Library of Congress: Washington 1913. ff.6. [66.]*

LIST of references on abandoned farms. Library of Congress: Division of bibliography. Washington 1913. pp.6. [83.]*

BRIEF list of references on state and municipal farm colonies. Library of Congress: [Washington] 1916. ff.3. [32.]*
— Additional references. 1916. ff.2. [15.]*

LIST of references on the mobilization of farm labor for war service. Library of Congress: [Washington] 1918. ff.9. [90.]*

ADOLPH WULFF, Bibliographia agrogeologica.

Landbouwhoogeschool: Mededeelingen (vol.xx): Wageningen 1921. p.284. [3302.]

MARGARET T[HOMPSON] OLCOTT and LOUISE O[LDMAN] BERCAW, Farm youth. A selected list of references to literature issued since January, 1920. Department of agriculture: Bureau of agricultural economics: Agricultural economic bibliography (no.17): Washington 1926. pp.[iv].42. [128.]*

— — [supplement]. Compiled by Esther M[arie] Colvin. . . . (no.65): 1936. pp.[i].v.196. [461.]

A[NNIE] M[URRAY] HANNAY, Control of production of agricultural products by governments. A selected bibliography. Department of agriculture: Bureau of agricultural economics: Agricultural economic bibliography (no.23): Washington 1927. pp.[iv].88. [274.]

MARGARET T[HOMPSON] OLCOTT, Large scale and corporation farming. A selected list of references. Department of agriculture: Bureau of agricultural economics: Agricultural economic bibliography (no.30): Washington 1929. pp.[ii].iv.87. [362.]*

LOUISE O[LDHAM] BERCAW, A. M. HANNAY and

## *Agriculture*

ESTHER M[ARIE] COLVIN, Bibliography on land settlement, with particular reference to small holdings and subsistence homesteads. Department of agriculture: Miscellaneous publications (no. 172): Washington 1934. pp.iv.492. [2701.]

EVERETT E[UGENE] EDWARDS, References on agricultural museums. Department of agriculture: Bibliographical contributions (no.29): Washington 1936. pp.v.43. [165.]*

WILLIAM A[NTON] NEWMAN and LORING K. MACY, Selected references on practices and use of labor on farms. Studies of changing techniques and employment in agriculture: Report (no.A–3): Philadelphia 1937. pp.495. [3960.]

ESTHER M[ARIE] COLVIN, Large scale and corporation farming. A selected list of references. Department of agriculture: Bureau of agricultural economics: Agricultural economic bibliography (no.69): Washington 1937. pp.[i].vii.121. [406.]*

CROP drying, with special reference to cereal and forage crops. References covering the period 1933–Aug. 1937. Science library. Bibliographical series (no.347): 1937. ff.8. [144.]

DERECHO rural. Universidad de Buenos Aires.

Facultad de derecho y ciencias sociales: Biblioteca: Serie bibliográfica (no.3): Buenos Aires 1938. ff.[i]. 58.10.3. [1000.]*

C. R. ENLOW, Review and discussion of literature pertinent to crop rotations for erodible soils. Department of agriculture: Circular (no.559): Washington 1939. pp.51. [158.]

AGRICULTURE in a war economy (April 1941–March 1942). Selected and annotated bibliography. Library of Congress: Legislative reference service: Bibliographies of the world at war (no.ii): Washington 1942. pp.81. [526.]

— Supplement . . . no.1. 1943. pp.106. [790.]

— Supplement . . . no.2. 1943. pp.[ii].64. [416.]

*no more published.*

EVERETT E[UGENE] EDWARDS, Selected references on Thomas Jefferson and his contributions to agriculture. Department of agriculture: Library: Library list (no.8): Washington 1944. pp.7. [100.]*

ANNIE M. HANNAY, Land settlement. A list of references. Department of agriculture: Library: Library list (no.9): Washington 1944. pp.167. [1375.]*

INDEX to maps and charts published by the Office of foreign agricultural relations. Department of agriculture: Washington 1950. pp.iv.40. [750.]*

A[NNESLEY] VOYSEY, Farming and kindred subjects. An annotated list of books for readers of eleven to fifteen years. School library association: 1951. pp.24. [135.]

EDOARDO VOLTERRA, Bibliografia di diritto agraria romano. Osservatorio italiano di diritto agrario: Firenze 1951. pp.141. [1500.]

LUCINDA CRILE, Selected bibliography on the origin and development of the farmers' cooperative demonstration work. Department of agriculture: Extension service circular (no.484): [Washington] 1952. pp.15. [100.]*

ÜBERSICHT über die agrarpolitische und agrarsoziologische aufsätze. Forschungsgesellschaft für agrarpolitik und agrarsoziologie: Bonn 1954 &c.*
*in progress?*

HARLEY H[ARRIS] BARTLETT, Fire in relation to primitive agriculture and grazing in the tropics: annotated bibliography. University of Michigan: Botanical gardens: Ann Arbor 1955–1957. pp. 570+[iii].873. [500.]*

AGRICULTURAL museums . . . 1943–1955. Science

library: Bibliographical series (no.740): 1956. ff.2. [40.]*

GEORGES PLAISANCE, Les formations végétales et paysages ruraux. Lexique et guide bibliographique. 1959. pp.3–424. [3500.]

HANS WERNER BAUM, Die sozialistische entwicklung auf dem lande im spiegel der erzählenden literatur. Zentralinstitut für bibliothekswesen: Abteilung aus- und weiterbildung: [Berlin 1959]. pp.32. [77.]

J. A. MCCORMICK, Radioisotopes in agriculture: animal husbandry, bacteriology, fertilizer uptake, plant physiology, photosynthesis, and entomology. A selected list of references. Atomic energy commission: Technical information service: Oak Ridge, Tenn. 1959. pp.67. [7500.]*

GERTY GÖRLITZ and HELMUT GÖRLITZ, Sozialistische umgestaltung in der landwirtschaft. Ein empfehlendes literaturverzeichnis. Deutsche bücherei: Sonderbibliographien (no.24): Leipzig 1960. pp.56. [200.]

— — 2. . . . auflage. 1962. pp.58. [250.]

BIBLIOGRAPHIE für die sozialistische landwirtschaft. (Bücher und zeitschriftenaufsätze . . . in auswahl). [Greifswald].

1955–1960. [By] Kurt Penzin [*and others*]. pp.256. [2790.]

1961. pp.262. [3016.]

PIERO FIORELLI, MAURO BANDINI and PAOLO GROSSI, Bibliografia del diritto agrario intermedio. Istituto di diritto agrario internazionale e comparato: Milano 1962– . pp.[ii].347+ . *in progress*.

**Forestry.**

[*bibliographies of agriculture and forestry are entered only under Agriculture.*]

1. *Periodicals*

HELEN MOORE, A union checklist of forestry serials. Department of agriculture: Forest service: [Washington 1936]. pp.viii.146. [2100.]*

FORESTRY abstracts coverage list, periodicals and series. Food and agriculture organization: FAO forestry and forest products studies (no.7): Rome 1953. pp.vii.180. [2250.]

PERIODICALS currently received. University of Natal: Wattle research institute: Library: [Pietermanitzburg 1958]. ff.4. [75.]*

LIST of periodicals and serial publications in the

library of the Department of forestry, university of Oxford. Second edition. [Oxford] 1961. ff.[i]. xvii.pp.284. [750.]*

LIST of periodicals in the Moon memorial library. Syracuse university: State university: College of forestry: [Syracuse] 1963. ff.[ii].84. [1000.]*

2. *General*

CHRISTOPH WILHELM JAKOB GATTERER, Allgemeines repertorium der forstwissenschaftlichen literatur. Ulm 1796. pp.xiv.288+[ii].iv.200. [2500.]

C[HRISTIAN] P[ETER] LAUROP, Handbuch der forst- und jagdliteratur. Zur forst- und jagdwissenschaft (vol.xv): Erfurt &c. 1830. pp.xix.444. [7500.]

— — 1829–1843. Frankfurt a. M. 1844. pp.xv. 180. [3250.]

CARL GEROLD, Die literatur der letzten zehn jahre aus dem gesammtgebiete der land- und forstwirthschaft . . . in deutscher, französischer und englischer sprache. Gerold's literarische handbibliothek (vol.ii): Wien [1865]. pp.[iii].239. [6000.]

EDUARD BALDAMUS, Die literarischen erscheinungen . . . auf dem gebiete der forst- und jagdwissenschaft. Prag.

1856–1865. pp.47. [400.]

1866–1870. Die literarischen erscheinungen . . . auf dem gebiete der forst-, haus- und landwirthschaft.

1871–1875.

*later issues are entered in the next section, under Germany.*

BARON FRIEDRICH VON LÖFFELHOLZ-COLBERG, Forstliche chrestomathie. Beitrag zu einer systematisch-kritischen nachweisung und beleuchtung der literatur der forstbetriebslehre und der dahin einschlagenden grund- und hülfswissenschaften. Berlin 1866–1874. pp.viii.168+viii.169–484+viii.485–859 + viii.859–1092 + viii.362 + [vi]. 159. [10,000.]

*vols.i–v.1 only; no more published.*

BARON FRIEDRICH VON LÖFFELHOLZ-COLBERG, Die bedeutung und wichtigkeit des waldes, ursachen und folgen der entwaldung, die wiederbewaldung . . . aus der einschlagenden litteratur systematisch und kritisch nachgewiesen und bearbeitet. Leipzig 1872. pp.vii.292. [500.]

J[ULIUS] T[HEODOR] C[HRISTIAN] RATZEBURG,

Forstwissenschaftliches schriftstellerlexikon. Berlin 1872. pp.xi.516. [3000.]
*reissued in 1874.*

F[RIEDRICH] W[ILHELM] SCHNEIDER, Katalog der forstakademischen bibliothek. Neustadt-Eberswalde 1872. pp.x.318. [2349.]

RICHARD [ALEXANDER] HESS, Lebensbilder hervorragender forstmänner und um das forstwesen verdienter mathematiker, naturforscher und nationalökonomen. Berlin 1885. pp.xxiv.440. [1500.]

LIST of books, &c., relating to botany and forestry, including the Cleghorn memorial library, in the library of the museum. Edinburgh museum of science and art: Edinburgh 1897. pp. iv.199. [3000.]

[WILLIAM PARKER CUTTER, *ed.*], List of publications relating to forestry in the department library. Department of agriculture: Library bulletin (no.24): Washington 1898. ff.2–5.7–93. [1237.]
— — [second edition]. Catalogue of publications [&c.]. [By Emma Beatrice Hawks] . . . (no.76): 1912. pp.302. [4001.]

KATALOG der bibliothek der Königlich sächsischen forstakademie, Tharandt. Tharandt 1900. pp.xvi.755. [12,047.]

CURRENT literature. Department of agriculture: Forest service: Library: Washington.

[i]. 1908. ff.[60]. [600.]*
[ii]. 1909. ff.[87]. [1000.]*
[iii]. 1910. ff.[25]*.pp.[34]. [1250.]
[iv]. 1911. pp.[48]. [1250.]
[v]. 1912. pp.[47]. [1250.]
[vi]. 1913. pp.[48]. [1250.]
[vii]. 1914. pp.[47]. [1250.]
[viii]. 1915. pp.[57]. [1500.]
[ix]. 1916. pp.[66]. [1500.]
[x]. 1917. pp.[57]. [1500.]
[xi]. 1918. pp.[56]. [1500.]
[xii]. 1919. pp.[38].[44]. [1500.]*
[xiii]. 1920. ff.108]. [2000.]*
[xiv]. 1921. ff.[101.] [2000.]*
[xv]. January–February 1922. pp.7.ff.8. [300.]*
[xvi]. April–December 1927. Compiled by Helen E. Stockbridge. ff.24.25.21.26. [1500.]*

[*continued as:*]

Forestry current literature.*

[xvii]. 1928. ff.25.25.30.27.23.24. [2500.]
[xviii]. 1929. ff.23.31.31. .24.28. [3000.]

[xix]. 1930. ff.26.24.25.31.22.24. [2500.]
[xx]. 1931. ff.27.21.23.19.21.23. [2000.]
[xxi]. 1932–1933. ff.(pp.)25.21.17.14.12.75.58. [4000.]

*new ser.*

i. 1934. pp.34.78.39.40.39.39. [3000.]
ii. 1935. pp.33. .47.51.24.44. [3000.]
iii. 1936. pp.45.49.39.35.67. [4000.]
iv. 1937. pp.488. [4000.]
v. 1938. ff.412. [3000.]
vi. 1939. ff.385. [3000.]
vii. January–April 1940. ff.

*no more published; the issues for March 1922–March 1927 form part of the* Journal of forestry (*Washington*).

TREES, forestry and lumbering. A list of books and of references to periodicals in the Brooklyn public library. Brooklyn 1911. pp.40. [300.]

ALFRED REHDER, The Bradley bibliography. A guide to the literature of the woody plants of the world published before the beginning of the twentieth century. Compiled . . . under the direction of Charles Sprague Sargent. Arnold arboretum: Publications (no.3): Cambridge [Mass.].

i–ii. Dendrology. 1911–1912. pp.xiv.566+viii.926. [75,000.]

iii. Arboriculture — economic properties of woody plants. 1915. pp.xii.806. [40,000.]

iv. Forestry. 1914. pp.xvi.589. [30,000.]

v. Index of authors and titles — subject index. 1918. pp.xxxii.1008.

TREES and forestry. A selected list of the more important books in the library. Carnegie library: Pittsburgh 1917. pp.18. [100.]

A[RTHUR] H[AROLD] UNWIN, The international forestry bibliography, including allied subjects. Critchmere 1921. pp.48. [6000.]

LIBRARY regulations and catalogue. Royal scottish arboricultural society: Edinburgh [printed] 1928. pp.14. [300.]

FOREST bibliography to 31st December, 1933. University of Oxford: Department of forestry: Oxford 1936–1938. pp.[iii].xviii.78+[iii].79–200 +[iv].201–274. [10,000.]
*incomplete; no more published.*

CURRENT monthly record of forest literature. Imperial forestry institute: Oxford.

1936. [Edited by P. S. Spokes]. ff.[186]. [3000.]

1937. ff.[157]. [2500.]
1938. ff.[189]. [2500.]

1939. ff.[246]. [3000.]

[*continued as:*]

Forestry abstracts.

i. 1939–1940. pp.[ii].8.319. [1500.]
ii. 1940–1941. pp.[ii].10.342. [1500.]
iii. 1941–1942. pp.[ii].10.360. [2000.]
iv. 1942–1943. pp.[ii].10.267. [2000.]
v. 1943–1944. pp.[ii].10.293. [2500.]
— Five year index. 1950– .
vi. 1944–1945. pp.[ii].24.273. [2000.]
vii. 1945–1946. pp.[ii].31.590. [2204.]
viii. 1946–1947. pp.[ii].36.600. [2857.]
ix. 1947–1948. pp.[ii].32.540. [2666.]
x. 1948–1949. pp.[ii].38.563. [3048.]
xi. 1949–1950. pp.[ii].40.591. [3460.]
xii. 1950–1951. pp.[ii].38.559. [3568.]
xiii. 1951–1952. pp.[ii].51.576. [4484.]
xiv. 1953. pp.[ii].49.516. [4050.]
xv. 1954. pp.[ii]. .542. [4434.]
xvi. 1955. pp.[ii].66.602. [4856.]
xvii. 1956. pp.[ii].64.614. [4659.]
xviii. 1957. pp.[ii].70.585. [4848.]
xix. 1958. pp.[ii].72.633. [5078.]
xx. 1959. pp.[ii].83.665. [5452.]
xxi. 1960. pp.[ii].82.692. [5237.]
xxii. 1961. pp.[ii].109.678. [5319.]
xxiii. 1962. pp.[ii].103.722. [6150.]

xxiv. 1963. pp.[vii].105.673. [5892.]
*in progress.*

THESES submitted for advanced degrees in the subject matter of forestry, in colleges and universities of the United States. Oregon state college library: [Corvallis] 1938. ff.[i].86. [840.]

FOREST RECREATION. A bibliography. Forest service: Library: Washington 1938. pp.vi.129. [929.]*

FRANZ GRÜNWOLDT, Die dokumentation in der forstwirtschaft. Neudamm 1940. pp.88. [1000.]

BIBLIOGRAPHIA forestalis. Centre international de sylviculture: Berlin.
1941. [Edited by] Josef Köstler. pp.276.

SOME reference books on domestic and foreign woods. Department of agriculture [&c.]: [Madison 1945]. pp.[i].13. [150.]*

T. D. A. library catalogue. Timber development association: Library: 1946.
— 4th edition. 1953. pp.69. [1600.]

BIBLIOGRAPHY of forestry and forest products. Reprinted . . . from . . . the Bibliography of agriculture of the United States Department of agriculture. [Food and agriculture organization: Washington].

1948. pp.[124]. [5000.]
1949. pp.[146]. [5000.]
1950.
1951. pp.216. [7500.]
1952. pp.iv.131+iv.118. [7500.]

*no more published.*

KATHERINE W[HIPPLE] HUGHES, RAY A. YODER and WILLIAM I. WEST, Forestry theses accepted by colleges and universities in the United States 1900–1952. Oregon state college: Bibliographic series (no.3): Corvallis 1953. pp.140. [2638.]

LIBRARY bulletin. University of Oxford: Department of forestry: Oxford 1953 &c.*

*in progress.*

LIBRARY catalogue of books, with lists of periodicals and Forestry commission publications. Forestry commission: 1957. pp.ii.134. [2000.]*

[DAGMAR FRANCISCI], Lesnictví. Universita: Knihovna: Výběrový seznam (nos.28, 33): Brně 1958–1959. pp.[viii].80+ff.[ii].15.3. [600.]*

DICTIONARY catalogue of the Yale forestry library, Henry S. Graves memorial library, Yale university. Boston 1962.

i. A–British F. pp.[xxviii].492. [10,322.]
ii. British G–Corp. pp.[ii].452. [9492.]
iii. Cors–Floods. pp.[ii].474. [9954.]
iv. Floors–France. pp.[ii].480. [10,080.]
v. Frank–Kis. pp.[ii].652. [13,692.]
vi. Kit–Mei. pp.[ii].432. [9072.]
vii. Mel–Park (state). pp.[ii].438. [9198.]
viii. Park (state A.–Res). pp.[ii].390. [8190.]
ix. Ret–Spro. pp.[ii].426. [8946.]
x. Spru–Trae. pp.[ii].322. [6762.]
xi. Trai–Wak. pp.[ii].464. [9744.]
xii. Wal–Z. Maps. pp.[ii].406. [8562.]

*this is a photographic reproduction of the catalogue cards; the figures in square brackets represent these cards.*

LIST of publications on the growth, structure, and identification of wood. Forest products laboratory: Madison 1963. pp.36. [400.]*

### 3. *Countries &c.*

#### *Africa*

JEAN HUBBARD, South african forestry since 1913. A bibliography. University of Cape Town: School of librarianship: Bibliographical series: [Capetown] 1949. ff.ix.26. [266.]*

A. L. GRIFFITH and B. E. ST L. STUART, An east african forest bibliography. East Africa high commission: East african agriculture and forestry research organization: [Nairobi 1955]. pp.122. [1500.]

*America, latin*

FRANCES J[OSEPHINE] FLICK, The forests of continental latin America (including european possessions). A bibliography of selected literature, 1920–1950. Department of agriculture: Bibliographical bulletin (no.18): Washington 1952. pp. v.193. [1968.]*

*America, north*

E[DWARD] N[ORFOLK] MUNNS, A selected bibliography of north american forestry. Department of agriculture: Miscellaneous publication (no. 364): Washington 1940. pp.ii.1142. [21,413.]

*Australia*

BIBLIOGRAPHY of australian timbers and other forest products. Council for scientific and industrial research: Division of forest products: South Melbourne.*

i. 1944. Compiled by M. I. Hulme. pp.66. [800.]
ii. 1945. pp.15. [150.]
iii. 1946–1947. pp.26. [250.]
iv. 1948–1949. pp.18. [200.]

## *Forestry*

### *Austria*

BARON ARTHUR [ADOLF SCHWÄGER] VON HOHENBRUCK, Oesterreichische land- und forstwirthschaftliche bibliographie. Archiv für landwirthschaft: Wiener landwirthschaftliche zeitung (vol. xxxv, beilage): Wien 1899. pp.ix.255. [7500.]

### *Belgium*

BIBLIOGRAPHIA universalis silviculturæ. Belgica. Service spécial belge des expériences et consultations scientifiques forestières: Groenendael.

1936. ff.[ii].34. [131.]
1937. ff.[i].34. [132.]
1938. ff.[i].33. [132.]
1939. ff.[i].38.44. [402.]
1940. ff.[i].17.29. [268.]
1941–1942. ff.[i].34.51. [420.]
1943–1945. ff.[i].24.54. [449.]

*no more published.*

### *Bulgaria*

LYUBA G. KHINCHEVA and ZLATI M. ZLATEV, Библиография на лесовъдската книжнина. Излѣзла на български отъ освобождението до 1937. София 1937. pp.128. [2000.]

## *Forestry*

### *Canada*

[see also *United States*]

LIST of publications of the Forest products laboratories of Canada, Ottawa, Vancouver. Department of northern affairs and national resources: Forestry branch: Ottawa 1958. pp.[ii].21. [300.]*

— Revised. 1962. pp.v.21. [300.]

### *Czechoslovakia*

[DAGMAR HOFMANOVÁ], Naše lesy. Bibliografie z fondů vědeskěho oddělení okresní knihovny. Českě Budějovicíce 1961. pp.40. [242.]

### *Denmark*

[GUSTAF] VILH[ELM] GRUNDTVIG, Danmarks forst- og jagtlitteratur. Fortegnelse over skrifter verdrørende skovbrug, hede- og klitdyrkning, jagt, jagtfauna og hunde. København 1901. pp. [iv].81. [675.]

DEN DANSKE skovbrugs-litteratur. Bibliographica universali silviculturae: Dania: København [ii: Springforbi].

i. –1925. Ved A[dolf] Oppermann, [Gustaf] V[ilhelm] Grundtvig. [1931–] 1935–1936.

pp.ix–lix.290+[ii].291–468+18. [4718.]
ii. 1926–1945. Ved Poul A. Hauberg. 1952–1957. pp.[iv].xix.250 + xxvii.255–507. [4030.]

*East Indies*

CATALOGUS van de bibliotheek van het boschwezen in Nederlandsch-Indië. 3[de] druck. Weltevreden 1910. pp.[iii].104. [1000.]
— Supplement. 1911. pp.[iii].62. [600.]

*Estonia*

V. HAINLA [*and others*], Metsanduslik kirjandu 1945–1958. Eesti NSV teaduste akadeemia: Zooloogia ja botaanika instituut: Tart 1960. pp.128. [750.]

*Finland*

BIBLIOGRAPHIA universalis silviculturæ. Fennia. Metsätieteellinen tutkimuslaitos [Helsingin yliopiston metsäkirjasto]: Helsinki.
1934. Toimittanut Eino Saari. ff.(pp.)120. [450.]
1935. ff.(pp.)170. [599.]
1936. ff.(pp.)105. [670.]
1937. ff.(pp.)105. [380.]
1938. ff.(pp.)83. [300.]
1939.
1940.

1941–1942. [*not published.*]

1943–1944. Toimittanut . . . Arvo Seppälä. 1959. ff.[vi].107. [1000.]*

*in progress.*

*France*

H. GRASSET, Lois forestières, ou analyse méthodique et raisonnée des lois, avis du conseil d'état, réglemens, arrêts et décisions concernant les eaux et forêts, la chasse et la pêche. Nyons 1826. pp. xiii.461. [581.]

ALFRED GERBAUT, Notice des principaux arrêts correctionnels rendus par la cour royale de Nanci en matière forestière depuis la promulgation du code. 1840. pp.xvi.331. [276.]

D. A. JACQUEMART, Bibliographie forestière française. 1851. pp.58. [623.]

*reprinted in 1852.*

CAMILLE VIGOUROUX, Bibliographie du reboisement. Clermont-Ferrand [printed] 1947. pp.14. [150.]

GEORGES PLAISANCE, Guide des forêts de France. [1961]. pp.413. [2000.]

*Germany*

FRIEDRICH BENEDICT WEBER, Handbibliothek der

deutschen forstwirthschaftlichen literatur; oder systematische anleitung zur kenntniss der deutschen forstschriften; von den ältesten bis auf die neuesten zeiten. Berlin 1803. pp.xxxii.375+x.116. [2500.]

THEODOR CHRISTIAN FRIEDRICH ENSLIN, Bibliothek der forst- und jagd-wissenschaft, oder verzeichniss aller brauchbaren . . . besonders aber vom jahr 1750 . . . in Deutschland erschienenen bücher, über alle theile des forst- und jagdwesens, über die fischerey und den vogelfang. Berlin 1823. pp.41. [1000.]

— — Zweite . . . auflage. Leipzig 1843. pp.[iii]. 102. [1750.]

F. W. SCHNEIDER, Bibliothek der forst- und jagd-literatur. Ein verzeichniss der . . . in Deutschland erschienenen schriften . . . 1842 bis . . . 1856. Berlin 1856. pp.iv.63. [1000.]

EDUARD BALDAMUS, Die erscheinungen der deutschen literatur auf dem gebiete der forst- und jagdwissenschaft, 1876–1880. Leipzig 1881. pp. [ii].32. [500.]

— — Die deutsche literatur der forst- und jagdwissenschaften, 1881–1885. pp.[ii].30. [500.]

*earlier issues are entered in the previous section.*

## *Forestry*

INTERNATIONALE bibliographie für forstwissenschaft in Bibliographia universali 634.9 F. Deutschland. Deutsche zentralstelle für forstliche bibliographie: Freiburg i. Br.*

1952. Bearbeiter: Helene Beyer. ff.[ii].x.246. [2000.]

### *Great Britain*

FORESTRY 1918–1946. [House of commons: Library:] Bibliography (no.6): [1946]. ff.6. [60.]*

R. ST. BARBE BAKER, Trees. National book league: Reader's guide: 1948. pp.12. [50.]

*the bibliography proper is by William Arthur Munford.*

### *Hungary*

AZ ORSZÁGOS erdészeti egyesület könyvtárának katalogusa. Budapest 1909. pp.122. [1750.]

### *India*

LIST of maps and publications relating to forest administration in the provinces under the government of India, sent to the Paris exhibition of 1878. Simla 1878. pp.[ii].12. [75.]

LIST of publications and maps relating to forest administration in India sent to the Edinburgh international forestry exhibition. Calcutta 1884. pp.[ii].26. [186.]

## *Forestry*

### *Poland*

MARIA PISARSKA, Bibliografia polskiego piśmiennictwa leśnego za pięciolecie 1945–1949. Państwowe wydawnictwo rolnicze i leśne: Warszawa 1951. pp.116. [2202.]

### *Romania*

BIBLIOGRAFIA forestieră romînă. Institutul de cercetări forestiere: Bucureşti.

1860–1956. [By] Teodor Bălănică [*and others*]. 1958. pp.534. [5867.]

1957. 1958. pp.59. [430.]

1958. 1960. pp.83. [592.]

*in progress.*

### *Russia*

F[EDOR] G[EORGEVICH] PEKHTEREV, Указатель литературы по лесной кооперации. Лесная кооперация (no.19): приложение: [Moscow 1922]. pp.32. [217.]

N. A. MASLOV, Лесная литература за 1930 и 1931 гг. Москва 1933. pp.100. [1592.]

ЛИТЕРАТУРА по лесной промышленности. Лесотехническая академия имени С. М. Кирова: Ленинград 1936 &c.

*in progress.*

## *Forestry*

INHALTSVERZEICHNISSE sowjetischer fachzeitschriften in deutscher übersetzung. Reihe IV. Land- und forstwirtschaft. Zentralstelle für wissenschaftliche literatur: Berlin 1952 &c.

*in progress.*

Z. L. VILYATSER, Библиографический справочник работ института 1932–1962. Всесоюзный научно-исследовательский институт лесоводства и механизации работ хозяйства: Пишкино 1962. pp.96. [2999.]

### *Scandinavia*

NORDISK skovbrugslitteratur. Nordisk skovunion: Helsingfors [&c.].

1947. Redaktion A. Howard Grøn.

1948. pp.84. [250.]

1949. pp.72. [125.]

[*continued as:*]

Nordisk skogsbruklitteratur. Nordisk skogsunion.

1950. Redaktion: Gustaf Sirén och Teppo Warras. pp.95. [200.]

1951. Redaktion: G. Sirén och Holger Mauria. pp.111. [300.]

1952. Redaktion: Bror-Anton Granvik. pp. 93. [300.]

1953. pp.89. [300.]

1954. pp.93. [300.]
1955. pp.86. [300.]
1956. pp.90. [300.]
1957. pp.69. [200.]
1958. pp.85. [250.]
1959. pp.79. [250.]

*in progress.*

*Spain*

JOSÉ JORDANA Y MORERA, Apuntes bibliográfico-forestales. Madrid 1873. pp.[vi].v.320. [1126.]

*Sweden*

DEN SVENSKA lantbrukslitteraturen. . . . Bibliografisk förteckning. Kungl. lantbruksakademi [Stockholm]: Uppsala.

i. Från äldsta tid t. o. m. år 1800. Uppgjord av P[er] M[agnus] Hebbe. 1939. pp.xvi.333. [2194.]

ii. Från 1801 till 1850. Utgiven av C[arl Algot Peter] Björkbom. 1945. pp.xvi.372. [2970.]

*Switzerland*

FORSTWESEN. Centralkommission für schweizerische landeskunde: Bibliographie der schweizerischen landeskunde (section v.9c[i]): Bern 1894. pp.x.150. [3000.]

— Supplement. . . . (v.9c.[i²]): 1907. pp.ix. 151–231. [1000.]

### *Temperate zone*

ALFRED REHDER, Bibliography of cultivated trees and shrubs hardy in the cooler temperate regions of the northern hemisphere. Harvard university: Arnold arboretum: Jamaica Plains, Mass. 1949. pp.xl.825. [40,000.]

### *Tropics*

PERIODICALS and books relating to tropical agriculture and forestry. Science library: Bibliographical series (no.75): [1933]. ff.[i].6. [98.]*

TROPICAL rain forests. Science library: Bibliographical series (no.681): 1949. ff.2. [37.]*

— Supplement. . . . (no.784): [1963]. pp.7. [145.]*

### *United States*

T. H. MCKEE, Reports of the Committee on agriculture and forestry, United States Senate, from the organization of the committee, March 4, 1863, to the close of the forty-ninth congress, 1887. Washington 1887. pp.6. [40.]

ERNEST BRUNCKEN, A bibliography of forestry in California. State board of forestry: Circular (no.3): Sacramento 1908. pp.16. [150.]

FOREST service bulletins, circulars, maps, and silvical leaflets relating to forests. Superintendent of documents: Price list (no.43): Washington [1909]. pp.20. [300.]
*frequently re-edited under various titles.*

JOHN M[ANVERS] BRISCOE, A classified list of american literature on forestry subjects. Revised. Maine forestry department: [Augusta] 1918. pp. 13. [125.]

LIST of publications. Syracuse university: New York state college of forestry: [Syracuse] 1929. pp.10. [200.]
— [another edition]. Publications. 1960. pp.12. [200.]*

INTERNATIONAL bibliography for forestry. United States of America. Internationaier [*sic*] verband forstlicher forschungsanstalten: [*s.l.*].*
1935. ff.[i].23. [175.]
1936. ff.[i].22. [175.]
1937. ff.[i].63. [188.]

ALF Z. NELSON, Rural assessment with special reference to forests. A bibliography (general and by states). Department of agriculture: Forest service: Washington 1940. ff.[i].29. [250.]*

## *Forestry*

ALF Z. NELSON, A selected bibliography on the economics of forestry in the United States. Department of agriculture: Forest service: Washington 1941. ff.v.172. [1772.]*

A LIST of the materials in the Austin Cary memorial forestry collection in the university of Florida. State library board: Tallahassee 1941. pp.[v].li.47. [433.]*

FORREST L. FOOR, Materials in the National archives relating to forest products. National archives: Reference information circular (no.19): [Washington] 1943. pp.8. [large number.]

ANNE M. AVAKIAN, Staff publications, California forest and range experiment station 1916–1949. Revised by D. H. Vinther. Berkeley 1949. pp. [ii].ii.48. [800.]*

FORESTRY and forest products: a select list of U.S. government publications. State library: Smithsonian bibliography (no.30): Pretoria 1951. pp.[ii].16. [220.]*

WILLIAM A[DAMS] DAYTON, United States tree books. A bibliography of tree identification. Department of agriculture: Bibliographical bulletin (no.20): [Washington] 1952. pp.[ii].32. [381.]

LIST of publications. Department of agriculture: Forest service: Forest products laboratory: Madison, Wis. 1952 &c.*

*in progress.*

LIST of publications for the years, 1938 thru 1952. Forest service: Pacific northwest forest and range experiment station: Portland, Ore. 1953. pp.26. [500.]*

BIBLIOGRAPHY of the Southeastern forest experiment station for the year. Department of agriculture: Forest service: Southeastern forest experiment station: Asheville 1953 &c.*

*in progress.*

CLODAUGH M. NEIDERHEISER, Forest history sources of the United States and Canada. A compilation of the manuscript sources. Forest history foundation: Saint Paul 1956. pp.xiii.140. [large number.]*

FOREST and range influences publications, 1905–1955. Department of agriculture: Forest service: Division of watershed management research: [Washington] 1956. pp.[ii].78. [900.]*

L[UCILLE] P. OLSEN and H[ELEN] A. WOODWORTH, Publications of the Lake states forest experiment station 1923–1955. Department of agriculture:

## *Forestry*

Forest service: Lake states forest experiment station: Station paper (no.39): [Washington] 1956. pp.130. [1250.]*

— — Supplement 1. Assembled by L. P. Olsen. 1961. pp.67. [600.]*

SOME references on forests and related natural resources. [Forest service: Washington 1958]. pp.16. [250.]*

MICHAEL M. REYNOLDS, Forests and forestry in West Virginia—a bibliography. West Virginia university: Library: Morgantown [1962]. pp. xiii.78. [1122.]

### *Yugoslavia*

ŠUMARSKA bibliografija . . . (1846–1946). Šumarska sekcija: Društva inženjera i tehničara N. R. Hrvatske: Zagreb 1947. pp.263. [7000.]

HAJRUDIN BUJUKALIĆ, Jedan prilog šumarskoj bibliografiji (1945–1953). Sarajevo 1953. pp.24. [300.]

## 4. *Physiology and pathology*

G[EORGE] F[ILIPPO] GRAVATT and S[ARAH] G. MANGANARO, List of translations pertaining to forest pathology. Department of agriculture: Division of forest pathology: Beltsvill, Md. 1949. pp.79. [800.]*

AGNES ELLIS MOORE, Bibliography of forest disease research in the Department of agriculture. Department of agriculture: Miscellaneous publication (no.725): Washington 1957. pp.ii.186. [2284.]*

THEODORE T[HOMAS] KOZLOWSKI, Tree physiology bibliography. Department of agriculture: Forest service: [Washington] 1958. pp.viii.316. [3000.]*

5. *Miscellaneous*

BIBLIOGRAPHY on forestry reconnaissance, photography, and forest fire protection by air. Air board: Intelligence branch: Search (no.1): Ottawa [1921]. ff.[6]. [16.]*

— — [another edition]. Selective bibliography [&c.]. 1922. ff.[10]. [31.]*

EFFECT of deforestation and afforestation on river floods. Science library: Bibliographical series (no.33): [1932]. ff.[i].2. [61.]*

EFFECT on fire on forests. A bibliography. Forest service library: Washington 1938. pp.vi.130. [605.]*

FRANCES J[OSEPHINE] FLICK and ELISABETH P[ATTERSON] BROWN, Economics of forestry. A bibliography for the United States and Canada,

1940–1947. Department of agriculture: Library list (no.52): Washington 1950. pp.vi.126. [3821.]

— — [supplement].

1948–1952. Compiled by F. J. Flick. . . . (no.52, supplement 1): 1955. pp.iv.136. [3466.]*

1953–1954. Compiled by Mary Jackman, Margaret W. Berry and Robert O. McMahon. . . . (no.52, supplement 2): 1959. pp.v.45. [1197.]*

FOREST fire protection [control] abstracts. Department of northern affairs and national resources: Forestry branch: Ottawa 1950 &c.*

*in progress.*

C. WILLIAM GARRARD, An annotated bibliography of aerial photographic applications to forestry. State university of New York: College of forestry: Bulletin (no.26): Syracuse 1951. pp. [vi].81. [500.]*

STEPHEN H[OPKINS] SPURR, Forest photogrammetry and aerial mapping. A bibliography 1887–1955. University of Michigan: School of natural resources: Ann Arbor 1956. ff.[i].60. [750.]*

## *Forestry*

G[EORGES] PLAISANCE, Bibliographie des sols forestiers. (Répertoire international). Centre d'études et de documentation paléontologiques: Annales (no.25): 1957. pp.165. [2250.]*

## Horticulture.



### 1. *Periodicals*

SANDRA SHEPARD, Bibliography of current horticultural periodicals. Cornell university: Bailey hortorium: Ithaca 1955. pp.[ii].54. [1150.]*

### 2. *History and early works*

W[ILLIAM] CAREW HAZLITT, Gleanings in old garden literature. Book-lover's library: 1887. pp. vii.263. [100.]

GEORGES GIBAULT, Étude sur la bibliographie et la littérature horticoles anciennes. 1905. pp.36. [150.]

CATALOGUE of early works on botany, agriculture and horticulture. Public library: Cardiff 1919. pp.24. [60.]

ELEANOUR SINCLAIR ROHDE, The old english

gardening books. New aldine library (vol.v): 1924. pp.xii.144. [125.]

LEGENDS and anecdotes of flowers: a list of references prior to 1892. Library of Congress: Washington 1948. ff.4. [38.]*

3. *General*

CATALOGUE of the library of the Pennsylvania horticultural society. Philadelphia 1840. pp.24. [200.]

LOUIS BOUCHARD [-HUZARD], Bibliographie. Ouvrages publiés jusqu'à ce jour sur les constructions rurales et sur la disposition des jardins. 1860. pp.[iv].826–876. [275.]

*50 copies printed.*

— — Deuxième édition. 1870. pp.[iv].947–1048. [500.]

*90 copies printed; this work also forms part of the author's* Traité des constructions rurales.

FRIEDR[ICH] JAC[OB] DOCHNAHL, Bibliotheca hortensis. Vollständige garten-bibliothek, oder alphabetisches verzeichniss aller bücher, welche . . . von 1750 bis 1860 in Deutschland erschienen sind. Nürnberg 1861. pp.lx.180. [3500.]

A CATALOGUE of the library of the Massachusetts horticultural society. Boston 1867. pp.65. [1500.]

— [another edition. By Mary Crane Hewett]. 1918–1920. pp.[ii].vii.372+x.373–587. [30,000.]

VERZEICHNISS der dem Gartenbau-vereine zu Darmstadt angehörigen bücher. Darmstadt [1875]. pp.31. [900.]

KATALOG der bibliothek des Vereins zur beförderung des gartenbaues in den königlich preussischen staaten. v. auflage. Berlin 1875. pp.[iii].79. [1500.]

— VI. auflage. 1897. pp.139. [2500.]

[J. GÉRAND], Catalogue des ouvrages contenus dans la bibliothèque de la Société d'horticulture de la Gironde. Bordeaux [printed] 1881. pp.16. [200.]

ALPHONSE LEFEBVRE, Catalogue de la bibliothèque de la Société d'horticulture de Picardie. Amiens 1884. pp.63. [425.]

RÈGLEMENT et catalogue de la bibliothèque. Société d'horticulture, d'arboriculture et de viticulture des Deux-Sèvres. Niort [printed] 1889. pp.44. [400.]

THE CODMAN collection of books on landscape gardening, also a list of books on trees and forestry. [Public library:] Boston 1899. pp.[v].26. [1250.]

G[EORGES] GIBAULT, Catalogue de la bibliothèque de la Société nationale d'horticulture de

France. 1900. pp.viii.156. [3500.]
— — Supplément. 1905. [350.]
— — 2[e] supplément. 1910. pp.iv.53. [1000.]

свод русской литературы по техникѣ садоводства. Российское общество садоводства. С.-Петербургъ 1899–1900. pp.224+440+446+454. [2500.]

C[HARLES] HARMAN PAYNE, The florist's bibliography. 1908. pp.80. [750.]
— — Second edition. 1913. pp.5–112. [900.]
— — Second supplement. 1924. pp.40. [400.]
*the first supplement forms part of the second edition.*

FLOWER gardens. A selected list of books. Public library: New York 1915. pp.12. [350.]

PATENTS for inventions. Fifty years subject index, 1861–1910. Class 6 (ii). Gardening and like appliances. Patent office: 1915. pp.10. [1000.]

A READING list on flower gardening. Public library: Special library list (no.8): Kansas City 1915. pp.11. [175.]

ABRIDGMENTS of specifications. Class 6 (ii). Gardening and like appliances. Patent office.
1909–1915. pp.xi[*sic*, ix].82. [300.]
1916–1920. pp.ix.52. [150.]

1921–1925. pp.ix.80. [250.]
1926–1930. pp.x.76. [250.]
*no more published.*

BRIEF list of books on gardening. Library of Congress: Washington 1917. ff.2. [20.]*

WAR gardens. Russell Sage foundation: Library: Bulletin (no.28): New York 1918. pp.4. [90.]

BIBLIOGRAPH [*sic*] on war gardens. Champion coated paper company: Champion monographs: [New York] 1918. pp.14. [40.]

MARY EVANS, Garden books, old and new. Pennsylvania horticultural society: [Philadelphia] 1926. pp.3–86. [600.]

[H. R. HUTCHINSON], The Lindley library. Catalogue of books, pamphlets, manuscripts and drawings. Royal horticultural society: 1927. pp.viii.488. [12,500.]

GARDENS & gardening. A selected list of books prepared by the Garden club of America [&c.] . . . to accompany an exhibition of garden books. Public library: New York 1927. pp.48. [600.]

CATALOGUE of books on gardening, poultry &

bees. Municipal libraries: Leicester 1927. pp.15. [gardening: 250.]

— [another edition]. 1933. pp.15. [250.]

BOOKS and pamphlets presented, purchased, or reviewed during the year 1927 [–1938] and deposited in the library. [Royal horticultural society: 1928–1939].

*issued half-yearly.*

A SELECTED list of english books for the amateur gardener, compiled by the . . . Royal horticultural society. [National book council:] Bibliography (no.108): [1929]. pp.8. [200.]

HORTICULTURAL abstracts. Imperial [*afterwards:* Commonwealth] bureau of fruit production [of horticulture and plantation crops]: East Malling.

i. 1931. pp.8.128. [428.]
ii. 1932. pp.12.180. [421.]
iii. 1933. pp.16.223. [616.]
iv. 1934. pp.20.244. [696.]
v. 1935. pp.20.288. [747.]
vi. 1936. pp.24.350. [1008.]
vii. 1937. pp.26.372. [1120.]
viii. 1938. pp.40.387. [1380.]
ix. 1939. pp.40.404. [1544.]
x. 1940. pp.38.404. [1589.]
xi. 1941. pp.33.364. [1525.]

xii. 1942. pp.27.229. [1584.]
xiii. 1943. pp.28.213. [1638.]
xiv. 1944. pp.37.263. [2030.]
xv. 1945. pp.36.273. [2082.]
xvi. 1946. pp.37.286. [2337.]
xvii. 1947. pp.51.326. [2808.]
xviii. 1948. pp.lxiii.333. [3162.]
xix. 1949. pp.cxvi.458. [3547.]
xx. 1950. pp.cxix.478. [3448.]
xxi. 1951. pp.cxlii.502. [4153.]
xxii. 1952. pp.cxxiii.651. [4568.]
xxiii. 1953. pp.cxxxvii.698. [4763.]
xxiv. 1954. pp.cxxxvi.633. [4493.]
xxv. 1955. pp.[ii].673. [4487.]
— Index . . . volumes XXI–XXV. Compiled by G. K. Argles. 1956. pp.[ii].iv.308.
xxvi. 1956. pp.cxliii.640. [4294.]
xxvii. 1957. pp.cxxxvi.641. [4029.]
xxviii. 1958. pp.cxliv.671. [4201.]
xxix. 1959. pp.cxliv.768. [4162.]
xxx. 1960. pp.[ii].815. [6332.]
— Index . . . volumes XXVI–XXX. 1961. pp. [iv].356.
xxxi. 1961. pp.clxx.936. [7295.]
xxxii. 1962. pp.clxxxviii.973. [7570.]
xxxiii. 1963. pp.clxxxi.829. [8330.]

*in progress.*

DOROTHY S[T. JOHN] MANKS, Selected catalog of the library. Massachusetts horticultural society: Boston 1932. pp.59. [1250.]

— — Revised edition. 1936. pp.64. [1250.]

— — — Supplement. 1942. pp.32. [750.]

VICTOR R[ICKMAN] BOSWELL, Lista selecta de publicaciones sobre hortalizas. Unión panamericana: Oficina de cooperación agrícola: Washington 1933. ff.[iii].81. [800.]*

GARDENS & gardening. Books in the Bristol public libraries. [Bristol 1933]. pp.40. [500.]

WILLIAM LANIER HUNT, The southern garden. University of South Carolina: Library extension publication (vol.i, no.1): Chapel Hill [1934]. pp. 57. [500.]

A SELECTED and classified list of books in english relating to agriculture, horticulture, etc. in the library of the Ministry of agriculture. Ministry of agriculture and fisheries: Bulletin (no.78): 1935. pp.vi.22. [500.]

— Third edition. 1954. pp.vi.89. [2000.]

L[EONARD] A[NTHONY] SPRINGER, Bibliographisch overzicht van geschriften, boek- en plaatwerken op het gebied der tuinkunst. Stichting 'Fonds landbouwexportbureau 1916–1918': Publicatie (no.15): Wageningen 1936. pp.166. [2750.]

DOROTHY S[T JOHN] MANKS, Selected catalog of the library. Revised edition. Massachusetts horticultural society: Boston 1936. pp.64. [1000.]
— — A supplement. 1942. pp.32. [500.]

READERS' guide to books on gardening. Library association: County libraries section [no.20]: 1939. pp.[iii].16. [400.]

[E. I. SAKS], Что читать по индивидуальному и коллективному огородничеству. Государственная библиотека СССР им. В. И. Ленина: Москва 1942. pp.8. [46.]

[WALTER HERBERT GAUMNITZ], Schools and war gardens. Some guides and resources. Office of education: Washington [1943]. pp.[iv].11. [50.]*

DIG for victory. Public libraries: Bristol 1943. pp.[4]. [93.]

LILI HEIMERS, Victory gardens for community, home, school. New Jersey state teachers college: Upper Montclair 1944. ff.[i].4. [50.]*

BRITISH gardens. A selected list of books, compiled by the Royal horticultural society. National book league: Book list (no.205): 1945. pp.6 (folder). [120.]

MAGDALENE R. NEWMAN and NELL W. SMALLWOOD, Victory gardening in the United States. Department of agriculture: Library: Library list (no.15): Washington 1945. pp.32. [450.]*

ARVID ØSTBY, Hedmarks-gardene i litteraturen. En bibliografi. Hedmark slektshistorielag: Hamar 1948. pp.28. [250.]

ALPINE garden society library. [Maidstone] 1948. pp.14. [150.]

— Fourth edition. Catalogue of the Alpine garden society library.... Compiled ... by K. G. Lazenby. Farnborough 1959. pp.22. [401.]

A. FRITZ, Catalogue bibliographique de l'art des jardins. Association belge des architectes de jardins: [Brussels 1947]. ff.[i].85. [1000.]*

WILFRID BLUNT, Flower books and their illustrators. An exhibition. National book league: [1950]. pp.58. [145.]

H. E. BATES, Flower gardening. National book league: Readers' guide: 1950. pp.24. [100.]

TORFINN SKARD, Norsk hagebruks bibliografi 1694–1944. Norske hageselskap: Oslo 1953. pp. 144. [2744.]

BRADFORD G. SEARS, Planning and construction references for landscape architects. State university of New York: College of forestry: Bulletin (no.30): [Syracuse 1953]. pp.vi.87. [500.]*

BOOKS for gardeners. Public libraries: Willesden 1953. pp.16. [300.]

F. S. GINZBURG, Больше плодов, ягод и винограда. Аннотированный указатель литературы. Центральная научная сельскохозяйственная библиотека: Москва 1956. pp.64. [300.]

GARDENS and the library. A selected list of materials found useful in Georgia libraries. [Atlanta] 1956. pp.14. [300.]*

A[LEKSANDR] E[VGENEVICH] CHZHAO, Украсим родину садами. Рекомендательный указатель литературы. Библиотека СССР имени В. И. Ленина: Москва 1956. pp.31. [40.]

SUBJECT index of papers published from East Malling research station in publications other than the annual reports from 1913 to 1953. Kent incorporated society for promoting experiments in horticulture: East Malling research station: [East Malling] 1957. pp.[iii].58. [800.]*

ABSTRACTS and index to the annual research reports. Kent incorporated society for promoting experiments in horticulture: East Malling research station.

1913–1953. 1957. pp.vi.175. [1000.]
1954–1961. 1963. pp.[iii].89. [500.]

BIBLIOGRAPHY on gardening found in the university of the Philippines libraries. Quezon City 1962. ff.[v].46. [374.]*

[PAUL LEFRANCQ, *ed.*], Florilège. Les livres et les fleurs. Exposition organisée à l'occasion des sixièmes floralies internationales. Bibliothèque municipale: Valenciennes 1962. pp.24. [51.]

DICTIONARY catalog of the library of the Massachusetts horticultural society. Boston 1963. pp. [vi].736+[ii].666+[ii].682. [43,764.]

*this is a photographic reproduction of the catalogue cards; the figure in square brackets represents these cards.*

4. *Miscellaneous*

PAUL KNUTH, Handbook of flower pollination, based upon Hermann Müller's work. . . . Translated by J[ames] R[ichard] Ainsworth Davis. . . . Volume I. Introduction and literature. Oxford 1906. pp.xix.383. [3860.]

JEAN [M. B.] MACDOWELL, Gardening for the

amateur in south Africa, south-west Africa and Rhodesia. A bibliography. University of Cape Town: School of librarianship: Bibliographical series: [Capetown] 1947 [1950]. ff.iv.23. [171.]*

ABSTRACTS on agricultural and horticultural engineering. Silsoe 1950 &c.

*details of this work are entered under Agriculture: Agricultural engineering, above.*

M[ENNICK] TRUMAN FOSSUM, Bibliography of marketing and other economic information for floriculture and ornamental horticulture. Agricultural marketing service: Washington 1954. pp.12. [150.]*

# *Special Subjects*

**Abortion.**

THE CONTROL of contagious abortion, 1934–38: Science library: Bibliographical series (no.437). 1938. ff.15. [271.]*

**Alfalfa.**

ALFALFA. Public library: Bulletin (no.80): Riverside, Cal. 1913. pp.47. [500.]

**Almond.**

ALMONDS and the almond industry. Science library: Bibliographical series (no.18): [1930]. ff.[i].4. [69.]*

**Anaemia.**

ANAEMIA in young pigs. Science library: Bibliographical series (no.432): 1938. ff.5. [91.]*

**Aphelinus mali.**

MERILL ARTHUR YOTHERS, An annotated bibliography on *Aphelinus mali* (Hald.), a parasite of the woolly apple aphid, 1851–1950. Department of agriculture: Bureau of entomology and plant quarantine (E-861): Washington 1953. pp.61. [310.]

**Aphis.**

CH[ARLES] DES MOULINS, Note bibliographique sur les pucerons. Bordeaux [1869]. pp.4. [15.]

**Apple.**

W. H. RAGAN, Nomenclature of the apple. A catalogue of the known varieties referred to in american publications from 1804 to 1904. Department of agriculture: Bureau of plant industry: Bulletin (no.56): Washington 1905. pp.383. [17,500.]

LOUISE O[LDHAM] BERCAW, The apple industry in the United States. Department of agriculture: Bureau of agricultural economics: Agricultural economic bibliography (no.19): Washington 1927. pp.[ii].ii.170. [808.]

GAS storage of apples. Science library: Bibliographical series (no.129): 1934. single sheet. [14.]*

**Areca nut.**

S.R.SRINIVASA IYENGAR [ṢRĪNĪVASA AIYAṄGĀR], List of references on coconut & arecanut. Agricultural college and research institute: Library service: Bibliographical series (no.3): [Madras] 1941. pp.[ii].10. [200.]*

**Arsenic.**

LIST of references on asenic. Library of Congress: [Washington] 1921. ff.8. [81.]*

R[OBERT] A[MBROSE] OSBORN and L[INNIE] A[RAH] PEASE, Selected references on the toxicity of arsenic. [Washington 1934]. pp.45. [450.]*

BIBLIOGRAPHY on the influence of arsenic and tin on the properties of iron and steel . . . 1864 to 1936. Iron and steel institute: Bibliographical series (no.4): 1937. pp.24. [100.]

TOXICITY to plants of arsenic in soil. Science library: Bibliographical series (no.753): 1957. ff.2. [50.]*

**Artichoke.**

BIBLIOGRAPHY of the industrial saccharification of artichokes. [Science library: Bibliographical series (no.41): 1932] single sheet. [12.]*

**Asclepiadaceae.**

A[NNA] GERALDINE WHITING, A summary of the literature on milkweeds (asclepias spp.) and their utilization. Department of agriculture: Bibliographical bulletin (no.2): Washington 1943. pp.[ii].41. [95.]

**Asparagus.**

ASPARAGUS [Asparagus officinalis), its culture, composition and botany. References covering the period 1920–May 1935. Science library: Bibliographical series (no.188): 1935. ff.[i].12. [277.]*

**Avocado.**

IRA J[UDSON] CONDIT, A bibliography on the avocado (*persea americana* Miller). University of California: Citrus experiment station: Riverside, Cal. [1939]. pp.[v].293. [4000.]*

— — Second edition. [1950.] pp.[iv].372. [5000.]*

**Bacon**

PRESERVATION of fresh pork and bacon. Science library: Bibliographical series (no.268): 1936. ff.2. [33.]*

**Bagasse.**

CLARENCE J[AY] WEST, The utilization of sugar cane bagasse for paper, boards, plastics, and chemicals. An annotated bibliography. Sugar research foundation: Scientific report series (no.3): New York 1946. pp.146. [408.]

**Bamboo.**

LIST of references on the industrial uses of bamboo, cane, rattan and reed. Library of Congress: Washington 1923. ff.3. [34.]*

**Banana.**

FLORENCE S[ELMA] HELLMAN, The banana: culture, industry, commerce. A bibliographical list. Library of Congress: [Washington] 1927. ff.26. [302.]*

L[OTTA] JEAN BOGERT, Dietary uses of the banana in health and disease. . . . Revised . . . edition. [United fruit company: Research department: New York 1942]. pp.67. [175.]

PUBLICATIONS on bananas. Imperial college on tropical agriculture: St Augustine, Trinidad 1955. pp.[4]. [50.]*

**Banana root weevil.**

MORTIMER D. LEONARD, A bibliography of the banana root weevil. San Juan, Porto Rico, 1931. pp.[ii].147–176. [250.]

**Barley.**

BARLEY breeding bibliography. Imperial bureau

of plant genetics: Cambridge 1930. ff.[ii].24. [150.]*

INFLUENCE of external conditions (soils, fertilizers &c.) on the nitrogen content of barley. Science library: Bibliographical series (no.194): 1935. ff.[i].4. [67.]*

— Supplement . . . (no.217): 1936. single sheet. [6.]*

**Bean.**

PHASEOLUS vulgaris (french bean). Public library of South Australia: Research service: [Adelaide] 1957. ff.16. [207.]

**Beet.**

PUBLICATIONS on sugarcane & beet. Imperial college of tropical agriculture: St Augustine, Trinidad 1955. pp.[17]. [150.]*

**Bloat.**

H. H. COLE [*and others*], A review of bloat in ruminants. National research council [:Reprint and circular series (no.124)]: Washington [1945]. pp.[ii].183–236. [222.]

**Boron.**

[E. M. GLYASS *and others*], Указатель литературы по бору. Академия наук СССР: Москва &c. 1937. pp.72. [1122.]

BORON as a plant nutrient. A bibliography of literature. American potash institute: Library: Washington.*

1936–1938. Prepared by Dorothy Harding. 1938. pp.[v].83.xvi. [400.]

1938. Compiled by Dorothy H. Jameson. 1939. pp.[iii].40.vi. [172.]

1939. [By] H. Jameson . . . and Catherine M. Schmidt. 1940. pp.[iii].81.xviii. [381.]

1940. 1941. pp.[v].68.xvi. [319.]

1941. [By] Cath. M. Schmidt [*and others*]. 1947. pp.[ii].61.xvii. [273.]

1953. Compiled by Helen N. Hudgins [*and others*]. 1955. pp.77.xxvi. [359.]

*no more published; not published for 1942–1952.*

EFFECTS of boron in iron and steel, 1946–48. Science library: Bibliographical series (no.684): 1949. ff.2. [27.]*

MARY E. SCHRODER and THOMAS C[EPHAS] GOODWIN, Boron, boron hybrides, and related substances. A bibliography. Library of Congress:

Technical information division: Washington 1955. pp.x.178. [514.]*

PURIFICATION of boron. Annotated bibliography based on soviet open literature. Library of Congress: Aerospace information division: AID report (no. B-63-114): [Washington] 1963. ff.iv.7. [6.]*

**Brucellosis.**

N. D. BEKLEMISHEV and I. K. KARAKULOV, Бруцеллез. Аннотированный указатель литературы (до 1951 г.). Академия наук Казахской ССР: Институт краевой патологии: Алма-Ата 1955. pp.509. [1850.]

[OTAKAR LÍBAL], Československá literatura o brucellose. Universita: Knihovna: Výběrový seznam (no.9): Brně 1955. ff.[ii].14. [133.]

**Butter oil.**

CARRIE B[ELLE] SHERFY, Bibliography on butter oil. Department of agriculture: Bibliographical bulletin (no.5): Washington 1944. pp.iv.40. [294.]

**Cacao.**

LIST of references on cacao (cocoa) and chocolate cultivation, manufacture and uses. Library of Congress: [Washington] 1921. ff.10. [123.]*

— Additional references [&c.]. 1933. ff.7. [86.]*

WITCH-BROOM diesease of cacao. Science library: Bibliographical series (no.264): 1936. ff.4. [74.]*

INSECTS infesting cacao beans, including their life history and methods of control. References covering the period 1930–1938. Science library: Bibliographical series (no.494): 1939. ff.5. [70.]*

WOLF MUELLER, Bibliographie des kakao, seiner geschichte, kultur, verwendung, verarbeitung, wirtschaftlichen bedeutung. Hamburg 1951. pp. [ii].120. [2700.]

PUBLICATIONS on cacao. Imperial college of tropical agriculture: St Augustine, Trinidad 1955. pp.[5]. [100.]*

BIBLIOGRAPHIE des travaux récents effectués sur le cacaoyer. Commission du Pacifique sud: Circulaire d'information technique (no.30): [Nouméa] 1957. pp.[i].33. [102.]*

**Cactus.**

SELECT list of publications in english on cacti in the Science library. Science library: Bibliographical series (no.47): [1932]. single sheet. [11.]*

**Casein.**

LIST of references on casein. Library of Congress: [Washington] 1923. ff.7. [67.]*

ARTIFICIAL wool from casein. Science library: Bibliographical series (no.312): 1937. single leaf. [11.]*

COMPOSITION and properties of casein, 1932–6. Science library: Bibliographical series (no.327): 1937. ff.8. [149.]*

**Cashew nut.**

CHEMISTRY and uses of cashew nut oil, 1941–5. Science library: Bibliographical series (no.627): ff.2. [40.]*

**Cereals.**

LODGING of cereals. Science library: Bibliographical series (no.225): 1936. ff.5. [80.]*

C[HARLES] H[AROLD] BRIGGS, A bibliography of cerealiana. Minneapolis 1938. pp.60. [4000.]

RUTH FLINN, Cereals in south Africa. A select bibliography. University of Cape Town: School of librarianship: Bibliographical series: [Capetown 1950]. ff.[v].20. [163.]

**Cheese.** [*see also* **Dairy farming.**]

THE MAKING of blue-veined cheeses. Science library: Bibliographical series (no.245): 1936. ff.4. [62.]*

— Supplement. . . . (no.375): 1938. ff.5. [83.]*

**Chestnut tree.**

R[OLLA] KENT BEATTIE, Bibliography of the chestnut tree blight fungus. Commission for the investigation and control of the chestnut tree blight disease in Pennsylvania: Harrisburg 1914. pp.[ii].32. [358.]

**Chewing gum.**

LIST of references on chicle and chewing gum. Library of Congress: [Washington] 1922. ff.4. [36.]*

**Chloropicrin.**

R[URIC] C[REEGAN] ROARK, Chloropicrin. Department of agriculture: Bureau of chemistry: Chemistry bibliography (no.1): Washington 1926. pp.73. [287.]*

— — Supplement. 1927. pp.8. [29.]*

— — Supplement. By W[ilbur] A[ugust] Gersdorff. Department of agriculture: Bureau of chemistry and soils: Insecticide division: 1930. pp.19. [80.]*

R. C. ROARK, A bibliography of chloropicrin, 1848–1932. Department of agriculture: Miscellaneous publications (no.176): Washington 1934. pp.88. [533.]

— — (First supplement . . .). [By] R. C. Roark and R. L. Busbey. 1935. pp.15. [81.]*

**Cinchona.**

CULTIVATION of cinchona spp. for the production of quinine. References covering the period, 1916–May 1939. Science library: Bibliographical series (no.483): 1939. ff.5. [100.]*

— Second edition. . . . 1916–1943. . . . (no.608): 1944. ff.7. [125.]*

R[EGINALD] E[RNEST] MOREAU, An annotated bibliography of cinchona-growing from 1883–1943. East african research institute, Amani: Nairobi 1945. pp.41. [700.]

FRANCES F. LOMBARD, Review of literature on

cinchona disease, injuries and fungi. Department of agriculture: Bibliographical bulletin (no.9): Washington 1947. pp.ii.70. [290.]

**Citric acid.**

CITRIC acid production by the mould aspergillus niger. Public library of South Australia: Research service: [Adelaide] 1955. ff.4. [33.]*

**Citrus.**

CITRUS fruits and their culture. Free public library: San Bernardino, Cal. 1913. pp.[24]. [400.]

SOILS of interest to California citrus growers. Public library: Bulletin (no.111): Riverside, Cal. 1914. pp.16. [200.]

LIST of references on citrus culture in Florida. Library of Congress: Washington 1914. ff.3. [28.]*

S. TOLKOWSKY, A bibliography of citrus literature. Jeffa [printed] [1929]. pp.8. [125.]

CITRUS. Science library: Bibliographical series (no.118): 1934. ff.[i].2. [34.]*

DEAN H[UMBOLDT] ROSE, HAROLD T. COOK and W. H. REDIT, Harvesting, handling, and transportation of citrus fruits. A digest . . . 1938 to 1948.

Department of agriculture: Bibliographical bulletin (no.13): Washington 1951. pp.178. [500.]

R. G. WEBB, The citrus fruit industry in south Africa. A bibliography. University of Cape Town: School of librarianship: Bibliographical series: [Capetown] 1951. ff.40. [184.]*

**Coconut.**

LIST of references on cocoanuts and cocoanut oil. Library of Congress: Washington 1916. ff.3. [28.]*

S. R. SRINIVASA IYENGAR [ṢRĪNĪVASA AIYAṄGĀR], List of references on coconut & arecanut. Agricultural college and research institute: Library service: Bibliographical series (no.3): [Madras] 1941. pp.[ii].10. [200.]*

F[RANK] E. PETERS, Bibliographie sur les aspects nutritifs de la noix de coco. Commission du Pacifique sud: Documentation technique (no.58): Nouméa 1954. pp.v.43. [181.]*

— — Document revu et corrigé (no.95): 1956. pp.v.49. [205.]*

**Cod-liver oil.**

JOHN RUEL MANNING, Bibliography on cod-liver

oil in animal feeding. Commissioner of fisheries: Report (1929, appendix = Bureau of fisheries document, no.1065): Washington 1929. pp.[ii]. 333–365. [250.]

**Coffee.**

LIST of references on the valorization of coffee. Library of Congress: [Washington] 1921. ff.6. [66.]*

H. R. M. DE HAAN, Overzicht der koffieliteratuur. Proefstation Malang: Mededeelingen (no.39): Soerabaja [printed] [1923]. pp.58. [678.]

*interleaved.*

BIBLIOGRAPHY of publications relating to coffee soils and fertilisers. Imperial bureau of soil science: Harpenden 1934. ff.v.55. [700.]*

ERNEST E. LOCKHART and FRED B. BLOOMHARDT, A survey of world literature on coffee. Coffee brewing institute: Publication (no.1): New York 1956. pp.[iii].23. [274.]

— — [another edition]. . . . (no.23): 1957. pp. [iii].32. [437.]

**Colza.**

[SIMONE DURANT and ALEXANDRA GATINEAU],

Le colza, ses produits et ses dérivés. Bibliographie rétrospective. Institut des corps gras: Service de documentation: Monographies: [1958.] pp.123. [500.]*

**Composts.**

THE INDORE process for the manufacture of composts. Science library: Bibliographical series (no.385): 1938. f.1. [14.]*

**Cotton, cottonseed.**

C[HARLES] J[EPHTHA] H[ILL] WOODBURY, Bibliography of the cotton manufacture. Waltham, Mass. 1901–1910. pp.213+53. [6000.]

LIST of references on cotton seed oil industry. Library of Congress: Washington 1907. ff.3. [22.]*

LIST of references on the cotton industry and trade. Library of Congress: Washington 1911. pp.42. [340.]*

— [supplement]. 1916. ff.29. [350.]*

LIST of references on cotton in Egypt and India. Library of Congress: Washington 1914. ff.4. [40.]*

LIST of references on the cost of manufacture of cotton goods and prices. Library of Congress:

Washington 1921. ff.5. [51.]*

SUMMARY of current literature. . . . Volume I [&c.]. British cotton industry research association: Information bureau: [*s.l.*] 1921 &c.

*in progress; vols.i–viii are reproduced from typewriting.*

EMMA B[EATRICE] HAWKS, Cotton: partial list of publications in english, exclusive of publications on disease and pests and publications of the state experiment stations. Agr[icultural] lib[rary] notes (vol.i, no.6, supplement): [Washington] 1926. ff.12. [350.]*

COTTON literature. Selected references. Compiled by Emily L[ouise] Day. Department of agriculture: Washington.*

i. 1931. pp.[ii].290. [2000.]
ii. 1932. pp.[ii].392. [2500.]
iii. 1933. pp.[ii].364. [2500.]
iv. 1934. pp.[ii].410. [2500.]
v. 1935. pp.[ii].518. [2500.]
vi. 1936. pp.[ii].628. [2500.]
vii. 1937. pp.[ii].742. [3000.]
viii. 1938. pp.[ii].903. [3909.]
ix. 1939. pp.[ii].939. [4099.]
x. 1940. pp.[ii].841. [3630.]
xi. 1941. pp.[ii].619. [3393.]

xii. January–June 1942. pp.[ii].249. [1514.]
*no more published.*

MILDRED C[ATHERINE] BENTON, Uses for cotton. Selected references in the english language. Department of agriculture: Bureau of agricultural economics: Agricultural economic bibliography (no.44): Washington 1932. pp.[ii].43. [266.]*

— — Supplement . . . 1933–July 1940. Compiled by Dorothy M[yrta] Ellis. . . . (no.91): 1941. pp.vi.129. [785.]

EMILY L[OUISE] DAY and RACHEL P[ENN] LANE, Economic development of the cotton-textile industry in the United States, 1910–1935. A selected bibliography. Department of agriculture: Bureau of agricultural economics: Agricultural economic bibliography (no.57): Washington 1935. pp.[ii].iv.137. [434.]*

MILDRED C[ATHERINE] BENTON, Financing american cotton production and marketing in the United States. Compiled . . . under the direction of Emily L. Day. Department of agriculture: Bureau of agricultural economics: Agricultural economics bibliography (no.61): Washington 1935. pp.ii.50. [290.]*

RACHEL P[ENN] LANE, Cotton and cottonseed.

A list of the publications of the United States department of agriculture on these subjects, including early reports of the United States patent office. Compiled . . . under the direction of Emily L[ouise] Day. Department of agriculture: Miscellaneous publication (no.203): Washington 1934. pp.iii.149. [917.]

EMILY L[OUISE] DAY, Government control of cotton production in the United States 1933–1935. A selected list of references. Department of agriculture: Bureau of agricultural economics: Agricultural economic bibliography (no.63): Washington 1936. pp.[iii].59. [326.]*

EMILY L[OUISE] DAY, Cotton linters. Selected references in english, 1900–July 1940. Compiled . . . under the direction of Mary G. Lacy. Department of agriculture: Bureau of agricultural economics: Agricultural economics bibliography (no.88): Washington 1940. pp.iii–vi.39. [205.]*

FRANCIS G. HICKMAN, *ed.* Cotton bibliography. . . . Selected readings for use in schools and libraries. Cotton trade journal: Education department: Memphis, Tenn. 1944. ff.60. [500.]*

CATALOGUE of educational material on cotton and cotton products. National cotton council of

America: Memphis [*c.*1945].
— Eighth revision. 1956. pp.[ii].31. [200.]*

COTTON (centralised buying) bill. [House of commons: Library:] Bibliography (no.12): [1946]. ff.5. [50.]*

BOOKS on spinning and weaving of cotton and rayons, mainly 1930–1946. Science library: Bibliographical series (no.637): 1947. ff.2. [32.]*

B. H. CAMINITA [*and others*], A review of the literature relating to affections of the respiratory tract in individuals exposed to cotton dust. Public health service: Public health bulletin (no.297): Washington 1947. pp.iii.86. [267.]

KARL L. ROLLINS, HAZEL H. FORT and DOROTHY B. SKAU, Cellulose-water relations: a selected bibliography with special reference to swelling of cotton and to its utilization in water-resistant fabrics. Department of agriculture: Library list (no.44): Washington 1949. pp.63. [523.]*

[L. D. HOWELL], Marketing cotton and cottonseed and their products. A preliminary list of references. Department of agriculture: Bureau of agricultural economics: Washington 1950. pp.42. [600.]*

R[OBERT] L[ANIER] KNIGHT, Abstract bibliography of cotton breeding and genetics, 1900–1950. Commonwealth bureau of plant breeding and genetics: Technical communication (no.17): Farnham Royal [1955]. pp.[v].256. [1191.]

ROSLYN KUPPERMAN RUBINS, KATHRYN L. BARINGER and DOROTHY B. SKAU, Improved cottonseed meals for feeding poultry and swine, 1954–55. An annotated bibliography. Department of agriculture: Agricultural research administration: Southern utilization research branch: Southern regional research laboratory: [New Orleans 1957]. pp.91. [159.]*

**Cream.**

RECONSTITUTED cream. Science library: Bibliographical series (no.428): 1938. ff.2. [20.]*

**Dairy farming.** [*see also* **Milk.**]

1. *Periodicals*

MURIEL F[ANNY] WRIGHT, Periodicals relating to dairying in the Department of agriculture. U.S. Department of agriculture: Bureau of agricultural economics: Agricultural economics bibliography (no.16): Washington 1926. pp.[ii].22. [200.]*

ESTHER M[ARIE] COLVIN, List of periodicals containing prices and other statistical and economic information on dairy products. Department of agriculture: Bureau of agricultural economics: Agricultural economics bibliography (no.71): Washington 1937. pp.[ii].114. [123.]*

2. *General*

ABRIDGMENTS of specifications relating to milking, churning, and cheese-making, A.D. 1777–1866, Comissioners of patents for inventions: 1874. pp.xxxix.72. [125.]

— 1876–1876. 1883. pp.vii.60. [100.]

*no more published.*

ABRIDGMENTS of specifications. Class 84. Milking [*afterwards:* Butter-making], churning, and cheese-making. Patent office.

1855–1866. pp.vi.24. [125.]
1867–1876. pp.vi.26. [150.]
1877–1883. pp.vi.26. [150.]
1884–1888. pp.vi.26. [150.]
1889–1892. pp.viii.36. [200.]
1893–1896. pp.viii.48. [250.]
1897–1900. pp.vii.36. [200.]
1901–1904. pp.vii.48. [200.]
1905–1908. pp.viii.58. [250.]

1909–1915. pp.vii.60. [250.]
1916–1920. pp.vi.22. [100.]

1921–1925. pp.vii.26. [100.]
1926–1930. pp.vii.42. [150.]
*no more published.*

PATENTS for inventions. Fifty years subject index, 1861–1910. Class 84. Milking, churning, and cheese-making. Patent office: 1914. pp.5. [450.]

R. B. VAN DER BERG, Milk and dairy science. Bibliography of scientific periodical literature (no.1): Wageningen 1927. ff.8.2. [75.]
*printed on one side of the leaf.*

SELECT BIBLIOGRAPHY on recent references on the nutrient requirements of dairy cows and the composition of feeding-stuffs. Science library: Bibliographical series (no.303): 1937. ff.11. [200.]*

THE DESIGN and operation of milking machines, 1900–1938. Science library: Bibliographical series (no.426): 1938. ff.10. [200.]*

DAIRY science abstracts. Imperial [Commonwealth] bureau of dairy science: Shinfield.
i. 1939–1940. pp.[vi].376.xv.xxvii. [2000.]

ii. 1940–1941. pp.[vi].334.xiv.xxv. [1750.]
iii. 1941–1942. pp.[vi].280.xiii.xxi. [1000.]
iv. 1942–1943. pp.[vi].204.xiii.xxi. [1000.]
v. 1943–1944. pp.208. [1250.]
vi. 1944–1945. pp.[vi].215.xii.xxviii. [1250.]
vii. 1945–1946. pp.[vi].297.xv.xxxix. [1500.]
viii. 1946–1947. pp.[vi].279.xv.xxxiv. [1500.]
ix. 1947–1948. pp.[vi].346.xliv. [2000.]
x. 1948–1949. pp [vi].220.xxxii. [1250.]
xi. 1949–1950. pp.[iii].250.xvi.xxxvii. [1500.]
xii. 1950–1951. pp.434. [2500.]
xiii. 1951–1952. pp.[vi].541.xxxvi.liv. [3000.]
xiv. 1952. pp.[vi].coll.966. pp.xviii.lix. [3000.]
xv. 1953. pp.[vi].coll.1012. pp.xvii.lix. [3000.]
xvi. 1954. pp.[vi].coll.1028 pp.xix.lxx. [3000.]
xvii. 1955. pp.[vi].coll.1076. pp.xix.lxxii. [3000.]
xviii. 1956. pp.[vi].coll.1072. pp.xix.lxi. [3000.]
xix. 1957. pp.[vi].coll.1056. pp.1057–1143. [3000.]
xx. 1958. pp.[vi].coll.1058.pp.1059–1141. [3000.]
xxi. 1959. pp.[vi].660. [3236.]

xxii. 1960. pp.[vi].748. [3604.]
xxiii. 1961. pp.[vi].715. [3546.]
xxiv. 1962. pp.[vi].728. [3697.]
xxv. 1963. pp.[vi].619. [3637.]
*in progress.*

LOUISE O. BERCAW, The dairy industry in the United States. Selected references on the economic aspects of the industry. Department of agriculture: Bureau of agricultural economics: Library list (no.11): Washington 1940. pp.59. [323.]*

— — 1940–1941. Compiled by Nellie G. Larson under the direction of Margaret T. Olcott. Department of agriculture: Bureau of agricultural economics: Agricultural economics bibliography (no.97): Washington 1942. pp.133. [485.]*

*the supplement is on green paper.*

LIST of publications, 1928–1944. Hannah dairy research institute: Kirkhill 1944. pp.16. [229.]

ALBERT R. CONLEY, History of research dealing with problems confronting the dairy industry which have been conducted by the agricultural experiment stations, 1881–1947. Bibliography of the research and general index of publications resulting from the research. Ohio state university: [Columbus] 1949. ff.ii.219. [4000.]*

EFFECTS of fats on animals, with special reference to the comparative value of butter and margarine (1933–1958). Science library: Bibliographical series (no.762): [1958]. pp.5. [100.]*

DAIRY marketing publications of the states and federal agencies associated with the North central regional committee on dairy marketing research, 1950–1961. [Urbana 1962]. ff.iii.64. [1000.]*

**Dates.**

DATE production in the United States: a selected list of references. Library of Congress: Washington 1946. ff.11. [86.]*

**Demand.**

BIBLIOGRAPHY on demand analysis and projections. Food and agricultural organization of the United nations: [Rome 1959]. pp.xv.167. [600.]*

**Derris.**

R[URIC] C[REEGAN] ROARK, A digest of the literature of derris (deguelia) species used as insecticides, 1747–1931. Department of agriculture: Miscellaneous publication (no.120): Washington 1932. pp.86. [456.]

R[URIC] C[REEGAN] ROARK, References to reviews and popular articles on derris. Bureau of entomology and plant quarantine: Division of insecticide investigations: Washington 1938. pp.28. [400.]

R[URIC] C[REEGAN] ROARK, Division of insecticide investigations abstracts of foreign and domestic patents relating to derris, lonchocarpus, tephrosia, and rotenone. Department of agriculture. Bureau of entomology and plant quarantine: Washington 1938. pp.69. [110.]*

INSECT pests of stored roots of derris and cubé (lonchocarpus). Science library: Bibliographical series (no.649): 1948. single leaf. [15.]*

**Dioscorea.**

YAMS (dioscorea spp.). References covering the period 1930–1948. Science library. Bibliographical series (no.680): 1949. ff.3. [45.]*

**Drainage, land.**

ELLEN A. HEDRICK, List of references to publications relating to irrigation and land drainage. Department of agriculture library: Bulletin (no. 41): Washington 1902. pp.181. [2000.]

SURFACE drainage. Science library: Bibliographical series (no.89): [1933]. ff.[i].2. [21.]*

G[ALINA] F[EDOROVNA] KRUPNOVA, Осушение и освоение заболоченных земель. Рекомендательный указатель литературы. Государственная... публичная библиотека имени М. Е. Салтыкова-Щедрина: Ленинград 1957. pp.32. [150.]

[VALENTINA AFANASEVNA ZVEREVA], Мелиорация болот и минеральных заболоченных земель. Библиографический указатель. Республиканский государственный институт по проектированию водохозяйственного и мелиоративного строительства [&c.]. Москва 1959. pp.132+156. [1750.]

**Eggs.**

THE DRYING of eggs. Science library: Bibliographical series (no.354): 1937. ff.3. [54.]*

**Ethylene.**

W[ILBUR] A[UGUST] GERSDORFF, Bibliography of ethylene dichloride. Department of agriculture: Miscellaneous publication (no.117): Washington 1932. pp.60. [469.]

H. D. YOUNG and R. L. BUSBEY, References to the use of ethylene oxide for pest control. Bureau of entomology and plant quarantine: Division of insecticide investigations: [Washington] 1935. pp. 16. [200.]*

BIBLIOGRAPHY of ethylenediaminetetraacetic acid. Bersworth chemical company: Framingham, Mass. 1953. pp.[iii].69. [605.]

BIBLIOGRAPHY on ethylene imine chemistry. Chemirad corporation and Hansborg & co.: Port Washington, N.Y. [1956]. ff.iv.65. [224.]*

SOME references of the use of ethylene oxide in fumigating foodstuffs . . . 1928–1956. Science library: Bibliographical series (no.767): 1959. pp.2. [50.]*

**Farms.**

[*only bibliographies on farm buildings are entered here; for farming* see **Agriculture.**]

VAJEN [EILLEEN] H[ITZ] FISCHER, The farmhouse. A list of references. Department of agriculture: Library list (no.19): Washington 1945. pp.124. [1122.]*

DIMITRIJ PUME and MARIE PUMEOVÁ, Zemědělské

stavitelství. Československá akademie zemědělských věd: Prameny literatury (2nd ser., no.3): [Prague 1957]. pp.76. [1252.]

STANLEY W. BROWN and VIRGIL E. BAUGH, Preliminary inventory of the records of the Farmers home administration. National archives: Preliminary inventories (no.118): Washington 1959. pp. vi.62. [large number.]*

**Fertilizers.**

STEPHEN CONRAD STUNTZ, Reference list on the electric fixation of atmospheric nitrogen and the use of calcium cyanamid and calcium nitrate on soils. Bureau of soils: Bulletin (no.63): Washington 1910. pp.89.

PUBLICATIONS relating to soils and fertilizers. List no.1[–77]. January–April, 1931[–December 1937]. Imperial bureau of soil science: Rothamsted experimental station: Harpenden 1931 &c.*

*the work is to be continued in a printed form.*

BIBLIOGRAPHY of soil science, fertilizers and general agronomy. Imperial [*afterwards:* Commonwealth] bureau of soil science: Harpenden.

1931–1934. pp.v–xxxi.475. [6000.]
1934–1937. pp.v–xv.540. [8000.]

1937–1940. pp.iii–xxiii.504. [5500.]
1940–1944.
1944–1947. pp.xxvii.451. [7500.]
1947–1950. pp.xxviii.535. [9000.]
1950–1953. pp.xxxviii.711. [10,000.]
1953–1956. pp.xxii.661. [12,000.]
1956–1959. pp.xxv.762. [13,000.]
*in progress.*

THE COMPARATIVE effects of different liming materials on soils and crops. Science library: Bibliographical series (no.341): 1937. ff.10 [150.]*

THE MECHANICAL application of fertilisers and manures. Science library: Bibliographical series (no.367): 1938. ff.12. [225.]*

THE USE of volcanic ash and basalt grit as fertilisers, 1922–1941. Science library: Bibliographical series (no.609): 1944. ff.2. [20.]*

NPK fertilizer abstract bulletin. British sulphur corporation. 1962 &c.*
*in progress.*

**Field crops.** [*see also* **Grass.**]

1. *Periodicals*

LIST of periodicals and reports examined by the

Commonwealth bureau of pastures and field crops.

1950. pp.72. [350.]
[1952]. pp.14. [400.]

## 2. *General*

HERBAGE abstracts. Imperial [*afterwards:* Commonwealth] bureau of plant genetics (herbage plants): Aberystwyth [vol.xx &c.: Farnham Royal].

i. 1931. pp.[ii].vii.106. [500.]
ii. 1932. pp.[iv].xii.268.65. [1250.]
iii. [1933]. pp.[iii].iii.290. [1500.]
iv. [1934]. pp.[iii].iv.318. [1500.]
v. [1935]. pp.[iv].ii.362. [2000.]
vi. 1936. pp.[iv].ii.489. [2500.]
vii. 1937. pp.[iv].440. [2500.]
viii. 1938. pp.[iv].483. [2085.]
ix. 1939. pp.[iv].400. [1776.]
x. 1940. pp.[ii].418. [1894.]
xi. 1941. pp.[iv].339. [1494.]
xii. 1942. pp.[iv].357. [2064.]
xiii. 1943. pp.[iv].430. [2054.]
xiv. 1944. pp.[iv].424. [1868.]
xv. 1945. pp.[iv].376. [1742.]
xvi. 1946. pp.[iv].389. [1899.]

xvii. 1947. pp.[iv].398. [1870.]
xviii. 1948. pp.[iv].313. [1241.]
xix. 1949. pp.[iv].302. [1285.]
xx. 1950. pp.[iv].268. [1441.]
xxi. 1951. pp.[iv].253. [1346.]
xxii. 1952. pp.[iv].308. [1399.]
xxiii. 1953. pp.[iv].277. [1312.]
xxiv. 1954. pp.[iii].303. [1277.]
xxv. 1955. pp.[iii].329. [1339.]
xxvi. 1956. pp.[iii].xxxi.318. [1363.]
xxvii. 1957. pp.vi.348. [1601.]
xxviii. 1958. pp.viii.xxx.348. [1672.]
xxix. 1959. pp.vii.356. [1831.]
xxx. 1960. pp.vii.416. [1926.]
xxxi. 1961. pp.vii.388. [2007.]
xxxii. 1962. pp.vii.390. [2151.]
xxxiii. 1963. pp.vii.322. [2073.]

*in progress.*

SELECT list of papers bearing on grassland establishment, management and improvement in the British Isles, 1930–36. Science library: Bibliographical series (no.280): 1935 [1936]. ff.10. [188.]*

LIST of reprints available on loan. Imperial bureau of pastures and forage crops: Aberystwyth.*

i. 1940. ff.[i].15. [116.]

LIST of translations filed. Imperial [Commonwealth] bureau of pastures and field crops: Aberystwyth.*

i–iii.
iv. 1945. pp.[ii].17.7. [200.]
v. 1948. pp.8. [69.]

FIELD crop abstracts. Commonwealth bureau of pastures and field crops: Aberystwyth.

i. 1948. pp.[iv].320. [1428.]
ii. 1949. pp.[iv].328. [1499.]
iii. 1950. pp.[iv].309. [1612.]
iv. 1951. pp.[iv].235. [1257.]
v. 1952. pp. [iii] 293. [1410.]
vi. 1953. pp.[iii] 331. [1613.]
vii. 1954. pp.[iii] 302. [1462.]
viii. 1955. pp.[iii].342. [1543.]
ix. 1956. pp.iv.314. [1432.]
x. 1957. pp.vi.330. [1690.]
xi. 1958. pp.vi.370. [1877.]
xii. 1959. pp.viii.380. [2093.]
xiii. 1960. pp.viii.382. [2070.]
xiv. 1961. pp.viii.352. [2041.]
xv. 1962. pp.viii.378. [2192.]
xvi. 1963. pp.viii.335. [2179.]

*in progress.*

R. R. HEAL and DONALD G. THOMPSON, Bibliographical report on herbicides. University of Minnesota: College of agriculture: Division of agronomy and plant genetics: [St Paul] 1949. pp. 68. [489.]*

CROPS and pastures: a select list of U.S. government publications. State library: Smithsonian bibliography (no.29): Pretoria 1951. ff.[ii].21. [320.]*

W[ALTER] F[REDERICK] JEPSON, A critical review of the world literature on the lepidopterous stalk borers of tropical graminaceous crops. Commonwealth institute of entomology: 1954. pp.vi.128. [500.]

DEAN E. MEDIN, References on methods of measuring production and utilization of range and pasture forage. A selected annotated list. Colorado department of game and fish: Technical bulletin (no.6): [Denver] 1960. pp.[ii].iii.43. [200.]*

JACK R[ODNEY] HARLAN, Bibliography of research in forage crop breeding conducted in southern states . . . 1951–1962. Oklahoma state university: Experiment station: Processed series (no.P–435): [Stillwater 1962]. ff.[13]. [200.]*

**Flax.**

LIST of references on the flaxseed industry. Library of Congress: [Washington] 1917. ff.3. [24.]*

LIST of references on the flax and linen industry. Library of Congress: [Washington] 1919. ff.15. [178.]*

ANTON BROSCH, Der flachs in der fachliteratur. Verband deutscher leinenindustrieller: Berlin 1922. pp.86.

CULTURE of flax. References covering the period 1933–Oct. 1937. Science library: Bibliographical series (no.352): 1937. ff.9. [150.]*

THE HISTORY of flax growing in England and Scotland. Science library: Bibliographical series (no.534): 1940. ff.2. [20.]*

**Fruit.**

1. *Periodicals*

PERIODICALS examined by the Imperial bureau of fruit production. East Malling 1931. ff.[36]. [149.]*

ESTHER M[ARIE] COLVIN, List of periodicals containing prices and other statistical and economic information on fruits, vegetables and nuts. Department of agriculture: Bureau of agricultural economics: Agricultural economics bibliography (no.55): Washington 1935. pp.[ii].238. [127.]*

2. *General*

[FRANÇOIS] THIÉRION, Revue bibliographique des principaux ouvrages français où il est traité de la taille des arbres fruitiers et particulièrement du pêcher. Troyes 1843. pp.108. [100.]

LIST of references on the California fruit growers' exchange. Library of Congress: Washington 1921. ff.2. [23.]*

KATHARINE G. RICE, Bibliography on the preservation of fruits and vegetables in transit and storage, with annotations. Department of agriculture: Library: Bibliographical contributions (no.4): Washington 1922. pp.[iii].76. [750.]*

FRUIT cultivation in deserts other than in the United States. Science library: Bibliographical series (no.56): [1932]. single sheet. [9.]*

MAMIE I[DELLA] HERB, Consumption of fruits and vegetables in the United States. An index to some sources of statistics. Department of agriculture: Bureau of agricultural economics: Agricultural economic bibliography (no.56): Washington 1935. pp.[ii].ii.125. [198.]*

WINTER washes for fruit trees with special reference to tar distillates. Science library: Bibliographical series (no.270): 1936. ff.12. [166.]*

G. K. ARGLES, A review of the literature on stock-scion incompatibility in fruit trees, with particular reference to pome and stone fruits. Imperial bureau of fruit production: Technical communication (no.9): East Malling 1937. pp.115. [194.]

THE DRYING of fruits and vegetables. References covering the period 1930–1937. Science library: Bibliographical series (no.380): 1938. ff.7. [115.]*
— 1930–1942 . . . (no.587): 1943. ff.10. [200.]*

IRRIGATION of small fruits. Science library: Bibliographical series (no.476): 1939. ff.3. [50.]*

VEGETABLE and fruit-growing. A small selection of the recently published books. Public libraries: Manchester 1944. pp.[6] (folder). [150.]

M[ARY] G. SMITH, Deciduous fruit. A bibliography. University of Cape Town: School of librarianship: Bibliographical series: [Capetown] 1947. ff.25. [168.]*
*limited to south african fruit.*

A[LEKSANDR] E[VGENEVICH] CHZHAO, Украсим родину садами. Рекомендательный указатель литературы. Государственная ... библиотека СССР им. В. И. Ленина: Москва 1956. pp.31. [40.]

VIKTORIYA ANDREEVNA GALUNSKAYA, За расширение садоводства в северо-западной зоне. Рекомендательный список литературы. Государственная... публичная библиотека имени М. Е. Салтыкова-Щедрина: Ленинград 1957. ff.20. [100.]*

ROBERT LANIER KNIGHT and ELIZABETH KEEP, Abstract bibliography of fruit breeding and genetics. Rubus and ribes a survey. Commonwealth bureau of horticulture and plantation crops: Technical communication (no.25): East Malling 1958. pp.254.

DAGMAR FRANCISCI, O pěstování zeleniny v zahrádkách. Universita: Knihovna: Výběrový seznam (no.32): Brně 1959. pp.[iv].26. [60.]*

**Fruit juice.**

FRUIT juices and vegetables, frozen. Public library of South Australia: Research service: [Adelaide] 1955. ff.12. [269.]*

**Germination.**

THE EFFECT of light on germination. Science library: Bibliographical series (no.700): 1950. ff.4. [66.]*

— Supplement. . . . (no.786): [1963]. pp.7. [150.]*

**Goat.**

EMMA B. HAWKS, Cattle, sheep and goat production in the range country. A selected list of publications issued by the U.S. department of agriculture and the agricultural colleges, experiment stations and departments of agriculture of the seventeen range states. Department of agriculture: Bibliographical contributions (no.19): Washington 1928. pp.78. [1000.]*

**Grain.**

INDEXES to subject matters of the reports of the House of commons. . . . Agriculture and corn trade, 1820–1834. 1834. pp.[vi].120. [5200.]

ABRIDGMENTS of specifications relating to grinding grain and dressing flour and meal, A.D. 1623–1866. Commissioners of patents for inventions: 1876. pp.xix.336. [500.]

ABRIDGMENTS of specifications. Class 48. Grain and seeds, treating (including flour and meal). Patent office.

1855–1866. pp.xi.98. [400.]
1867–1876. pp.xi.110. [400.]
1877–1883. pp.x.70. [300.]
1884–1888. pp.x.72. [300.]
1889–1892. pp.x.62. [250.]
1893–1896. pp.x.68. [250.]
1897–1900. pp.x.54. [200.]
1901–1904. pp.xii.78. [300.]
1905–1908. pp.x.82. [300.]
1909–1915. pp.ix.123. [500.]
1916–1920. pp.ix.55. [200.]
1921–1925. pp.xi.70. [250.]
1926–1930. pp.x.74. [250.]
*no more published.*

ABRIDGMENTS of specifications relating to agriculture. Division II. Barn and farmyard implements (including the cleaning, drying, and storing of grain), A.D. 1636–1866. 1879. pp.xv.480. [750.]

— 1867–1876. 1879. pp.xii.252. [500.]

SELECT list of references on grain, wheat, flour, etc. Library of Congress: Washington 1911. ff.64. [404.]*

PATENTS for inventions. Fifty years subject index, 1861–1910. Class 58. Grain and seeds, treating (including flour and meal). Patent office: 1914. pp.11. [2000.]

LIST of references on the grain growing possibilities of northern Africa, Asia Minor and Mesopotamia. Library of Congress: [Washington] 1917. ff.3. [15.]*

LIST of references on the marketing of grain and grain exchanges. Library of Congress: [Washington] 1921. ff.12. [151.]*

SPROUTED grain as feed for livestock. Science library: Bibliographical series (no.246): 1936. ff.3. [50.]*

I[SABELLE] H[ENRIETTE] VAN EEGHEN, Inventarissen van de archieven van de Directie van de moscovische handel, Directie van de oorstersche [*sic*] handel en reederijen, Commissarissen tot de graanhandel en Commissie voor de graanhandel. Gemeentelijke archiefdienst: Amsterdam 1961. pp.104. [10,000.]

**Grain elevators.**

LIST of references on government owned terminal grain elevators. Library of Congress: Washington 1915. single leaf. [9.]*

LIST of references on farmers' elevators: coöperative, &c. Library of Congress: [Washington] 1922. ff.5. [65.]*

**Grass.** [*see also* **Field crops.**]

HELEN M. DRUMMOND and E. H. LEATHAM, Bibliography of New Zealand tussock grasslands. Department of scientific and industrial research: Bulletin (no.132): Wellington 1959. pp.41. [400.]

AGNES CHASE and CORNELIA D. NILES, Index to grass species, Smithsonian institution. Boston 1963.

i. A–D. pp.[iv].608. [20,800.]
ii. E–Pan. pp.[ii].598. [20,800.]
iii. Pap–Z. pp.[ii].538. [20,800.]

*this is a photographic reproduction of cards; the figures in square brackets represent these cards.*

**Ground-nut.**

LIST of references on peanut meal or flour. Library of Congress: Washington 1919. single leaf. [10.]*

LIST of references on the peanut industry. Library of Congress: Washington 1923. ff.12. [165.]*

HELEN E[MMA] HENNEFRUND, The peanut industry. A selected list of references on the economic aspects of the industry, 1920–1939. Department of agriculture: Bureau of agricultural economics: Agricultural economic bibliography (no.80): Washington 1939. pp.viii.238. [641.]*

FOOD preparation from ground-nut cake. Science library: Bibliographical series (no.478): 1939. ff.2. [27.]*

— Supplement. . . . (no.638): 1947. ff.2. [23.]*

INSECT pests of stored peanuts. Science library: Bibliographical series (no.648): 1948. ff.3. [40.]*

**Guar.**

THEODORE HYMOWITZ, Annotated bibliography on the dieseases of guar. Oklahoma state university: Experiment stations: Processed series (no. P-468): [Stillwater] 1963. ff.23. [122.]*

**Guayule.**

ALAN J[OHNSTON] BLANCHARD, Guayule. A list of references. Department of agriculture: Soil conservation service: Soil conservation bibliography (no.4): Washington 1942. pp.54. [225.]*

— — Revised by Anne Avakian and Ruby W[ilson] Moats. Department of agriculture: Library: Library list (no.10): 1944. pp.[ii].iii.61. [475.]*

**Hannah dairy research institute.**

LIST of publications, 1928–1944. Hannah dairy research institute: Kirkhill 1944. pp.16. [229.]

**Hay.**

J[OHN] W[ILLIS] WEAVER and GEORGE E. ZERFOSS, Natural and artificial drying of hay. Selected references. Tennessee valley authority: Commerce department: Agricultural engineering development division: [Knoxville, Tenn.] 1940. ff.[ii].43[*sic*, 45]. [533.]*

MECHANICAL drying of hay. (References mainly 1946 to October 1947). Science library: Bibliographical series (no.645): 1947. ff.6. [100.]*

**Hazelnuts.**

BIBLIOGRAPHY of hazelnuts and the hazelnut industry. Science library: Bibliographical series (no.17): [1930]. ff.[i].2. [36.]*

**Helminthology.**

ADOLPH MODEER, Bibliotheca helminthologica seu enumeratio auctorum qui de vermibus scilicet eryptozoois, gymnodelis, testaceis atque phytozoois tam vivis quam petrificatis scripserunt. Erlangae 1786. pp.[viii].223. [2000.]

CORRADO PARONA, Elmintologia italiana (bibliografia-sistematica-storia). Pavia 1889. pp.12. [160.]

J[OHANN] CH[RISTOPH] HUBER, Bibliographie der klinischen helminthologie. München [1891–] 1895. pp.[iii].381. [6000.]

— — Supplementheft. Jena 1898. pp.[ii].22. [300.]

BIBLIOGRAPHY of helminthology. Imperial bureau of agricultural parasitology: St. Albans.

1930. Compiled by A[rthur] Walton. 1932–1933. pp.70. [1000.]
— Supplement. 1933. pp.[ii].18. [150.]
1931. 1932–1933. pp.[ii].93. [1500.]
1932. 1934. pp.[ii].97. [1500.]
1933. 1935. pp.[ii].101. [1500.]

*amalgamated with* Helminthological abstracts.

HELMINTHOLOGICAL abstracts. . . . A résumé of

the current periodical literature. Institute of agricultural parasitology [1936 &c. Imperial bureau of agricultural parasitology]: St. Albans.

i. 1932–1933. Edited by R. T. Leiper. pp. [iv].282. [442.]

ii. 1933–1935. Edited by R. T. Leiper and B. G. Peters. pp.[ix].270. [493.]

iii. 1934–1936. pp.[ii].280. [737.]

iv. 1935–1937. pp.[ii].276. [727.]

v. 1936–1938. pp.[ii].273. [718.]

vi. 1937–1939. pp.[ii].309. [732.]

vii. 1938–1940. pp.[ii].288. [726.]

viii. 1939–1941. pp.[iii].258. [665.]

ix. 1940–1942. pp.[ii].194. [577.]

x. 1941–1943. pp.[ii].168. [542.]

xi. 1942–1944. pp.[ii].133. [464.]

xii. 1943–1945. pp.[ii].130. [451.]

xiii. 1944. 1944–1947. pp.[ii].145. [519.]

xiv. 1945. 1945–1948. pp.[ii].168. [556.]

xv. 1946. 1946[–1950]. pp.[ii].205. [659.]

xvi. 1947. 1948[–1951]. pp.[ii].220. [700.]

xvii. 1948. 1948[–1952]. pp.[ii].246. [797.]

xviii. 1949. 1949[–1952]. pp.[ii].318. [874.]

xix. 1950. 1950[–1953]. pp.[ii].369. [914.]

xx. 1951. pp.[ii].402. [1018.]

xxi. 1952. pp.[ii].398. [1010.]

xxii. 1953. pp.[ii].478. [1012.]

xxiii. 1954. pp.[ii].480. [1001.]
xxiv. 1955. pp.[ii].526. [1011.]
xxv. 1956. pp.[ii].544. [939.]
xxvi. 1957. pp.415. [631.]
xxvii. 1958. pp.207. [391.]
xxviii. 1959. pp.
xxix. 1960. pp.[ii].616. [2332.]
xxx. 1961. pp.iv.650. [2800.]
xxxi. 1962. pp.[ii].648. [3002.]
xxxii. 1963. pp.[iv].507. [3087.]

*in progress.*

MAY R. YOUNG, Helminth parasites of New Zealand. A bibliography. Imperial bureau of agricultural parasitology (helminthology): St. Albans 1938. pp.19. [75.]

MAY R. YOUNG, Helminth parasites of Australia. A bibliography. Imperial bureau of agricultural parasitology (helminthology): St. Albans 1939. pp.[ii].145. [1000.]

BIBLIOGRAPHY of phenothiazine as an anthelminthic. Imperial bureau of agricultural parasitology: St. Albans 1942. pp.[ii].6. [120.]

B[ERND] HÖRNING, Die sovjetische helminthologie (literaturübersicht der jahre 1946–1953). Freie universität Berlin: Osteuropa-institut: Reihe medizin: Berichte (no.30): Berlin 1957. pp.99. [700.]*

T. E. GIBSON, Veterinary anthelmintic medication. A review of the literature. Commonwealth bureau of helminthology: Technical communication (no.33): Farnham Royal 1962. pp.[v].172. [1500.]

**Hemp**

LEWIS P[AUL] MCCANN, Kenaf (Hibiscus cannabinus L.). A bibliographical survey. Department of agriculture: Bibliographical bulletin (no.17): Washington 1952. pp.20. [115.]

**Hessian fly.**

J. S. WADE, An annotated bibliography of the hessian fly, phytophaga destructor (Say). Department of agriculture: Miscellaneous publication (no.198): Washington 1934. pp.100. [1256.]

**Heterodera.**

THE ROOT-INFESTING eelworms of the genus heterodera. A bibliography and host list. Imperial bureau of agricultural parasitology: St. Albans 1931. pp.viii.99. [800.]

**Honey.**

ANTISEPTIC and bacteriostatic properties of honey. Science library: Bibliographical series (no.667): 1949. pp.2. [20.]*

**Insemination, artificial.**

ERNEST HENRY BREUER, Artificial insemination. New York 1948. ff.x.47. [173.]*

**International institute of agriculture.**

[MARA CAMERANI-TEODOROVA], Bibliographie concernant l'histoire et l'activité de l'institut international d'agriculture (1905–1945). Rome 1945. pp.3–38. [326.]

**Iodine.**

IODINE in the animal world. (Farm animals). Science library: Bibliographical series (no.345): 1937. ff.11. [149.]*

— II. The iodine content of foodstuffs, 1925–37. . . . (no.382): 1938. ff.13. [260.]*

BIBLIOGRAPHY of the literature on sodium and iodine in relation to plant and animal nutrition.

Chilean nitrate educational bureau: New York. i. 1948. pp.[ii].123. [1000.]
*no more published.*

IODINE content of foods. Annotated bibliography, 1825–1951. Chilean iodine educational bureau: 1952. pp.vii.183. [326.]

**Irrigation.**

ELLEN A. HEDRICK, List of references to publications relating to irrigation and land drainage. Department of agriculture: Library bulletin (no.41): Washington 1902. pp.181. [1778.]

SELECT list of references on pumping for irrigation. Library of Congress: Washington 1908. ff.5. [19.]*

ORGANIZATION, work, and publications of irrigation investigations. Office of experiment stations: Washington 1909. pp.12. [60.]

GEORGE HARTWELL GIFFORD, List of references to publications relating to irrigation in the Public library of New South Wales. Sydney 1913. pp. [iii].32. [450.]

BRIEF list of references on irrigation. Library of Congress: Washington 1914. ff.6. [55.]*

LIST of references on irrigation and irrigation projects in Australia, Canada, England, India, Japan and south Africa. Library of Congress: [Washington] 1924. ff.12. [110.]*

IRRIGATION, overhead. Public library of South Australia: Research service: [Adelaide] 1957. ff.15. [261.]*

ELIZABETH GOULD DAVIS, EDNA S. SUBER and WAYNE F. EHLERS, The economics of irrigation in the eastern United States. A list of annotated references. Department of agriculture: Library list (no.66): Washington 1958. pp.68. [682.]*

**Kapok.**

A LIST of references on kapok. Library of Congress: [Washington] 1924. ff.5. [44.]*

CHEMICAL composition of the products of the kapok tree. Science library: Bibliographical series (no.460): 1939. ff.2. [34.]*

**Lactobacillus acidophilus.**

W[ILLIAM] D[ODGE] FROST and HAZEL HANKINSON, Lactobacillus acidophilus. An annotated bibliography. Milton, Wis. [1931]. pp.160. [174.]

**Livestock.**

A. MAYGRIER, Essai d'une bibliographie sommaire et raisonnée de la vaccine ou revue rapide des principaux écrits, travaux et discussions sur cette maladie depuis Jenner jusqu'aux dernières expériences de mm. Lafosse, H. Bouleye et Chauveau. 1865. pp.viii.28. [150.]

W. DIECKERHOFF, Geschichte der rinderpest und ihrer literatur. Berlin 1890. pp.vii.270. [250.]

JEAN DE CLAYBROOKE and [ ] WUNDERLICH, Liste des principaux ouvrages français et étrangers traitant des animaux de basse-cour. Société nationale d'acclimatation: [1895]. pp.38. [350.]

GEORGE FAYETTE THOMPSON, Index to literature relating to animal industry in the publications of the Department of agriculture, 1837 to 1898. Department of agriculture: Division of publications: Bulletin (no.5): Washington 1900. pp.676. [25,000.]

SELECT list of references on cattle ranches, fencing, etc. Library of Congress: Washington 1907. ff.13. [94.]*

— Additional references. 1911. single leaf. [8.]*

PATENTS for inventions. Fifty years subject

index, 1861–1910. Class 5. Patent office: 1913.
i. Farmyard and like appliances (other than housing and feeding animals). pp.8. [1000.]
ii. Housing and feeding animals. pp.7. [800.]

ABRIDGMENTS of specifications. Class 5. Patent office.
i. Farmyard and like appliances (other than housing and feeding animals).
1909–1915. pp.viii.96. [300.]
1916–1920. pp.vii.36. [100.]
1921–1925. pp.viii.72. [200.]
1926–1930. pp.viii.77. [250.]
ii. Housing and feeding animals.
1909–1915. pp.viii.99. [300.]
1916–1920. pp.vii.45. [150.]
1921–1925. pp.ix.95. [300.]
1926–1930. pp.ix.80. [250.]
*no more published.*

[V. R. Y. CROESEN], Literatuur over rundveteelt, voeding, en verpleging geschreven door Nederlanders. Koninklijke nederlandsche landbouwvereeniging: Mededeelingen en berichten (no.12): 's-Gravenhage 1916.

LIST of references on live stock exchanges. Library of Congress: [Washington] 1920. ff.2. [16.]

BRIEF list of references on the live stock industry in the tropics. Library of Congress: [Washington] 1921. ff.4. [40.]*

KATHARINE JACOBS, Livestock financing. A selected list of references relating to the financing of the livestock industry in the United States. Department of agriculture: Bureau of agricultural economics: Agricultural economics bibliography (no.7): Washington 1925. pp.[ii].28. [168.]*

— — [another edition]. . . . (no.62): 1935. pp. [ii].62. [362.]

*limited to the publications of 1915–1935.*

EMMA B. HAWKS, Cattle, sheep and goat production in the range country. A selected list of publications issued by the U.S. Department of agriculture and the agricultural colleges, experiment stations and departments of agriculture of the seventeen range states. Department of agriculture: Bibliographical contributions (no.19): Washington 1928. pp.78. [1000.]*

TREATMENT of mastitis of cows. Science library: Bibliographical series (no.250): 1936. ff.3. [37.]*

DISEASES and disorders of livestock caused by mineral deficiency or excess. Science library: Bibliographical series (no.297): 1937. ff.15. [300.]*

BLOOD analysis as an index of aphosphorosis and alcalcicosis in farm animals. Science library: Bibliographical series (no.390): 1938. ff.4. [63.]*

PEDRO GRASES, Contribución a la bibliografía venezolana de temas agropecuarios. Ministerio de agricultura y cría: Publicaciones: Carácas 1943. pp.3–180. [469.]

[E. I. SAKS], что читать о мерах увеличения поголовья скота и повышения его продуктивности. Краткий аннотированный список литературы. Государственная библиотека СССР имени В. И. Ленина: Москва 1943. pp.12. [100.]

РЕФЕРАТИВНО-БИБЛИОГРАФИЧЕСКИЙ справочник по животноводству. Москва.
1949.
1950.
1951. Под редакцией Г. Н. Доброхотова. 1953. pp.472. [1320.]
1952. 1956. pp.391. [1598.]

CORABEL BIEN and MARIE GOULD, Tropical beef cattle industry in the western hemisphere. Department of agriculture: Bibliographical bulletin (no.19): Washington 1952. pp.vi.157. [1749.]*

DUŠAN PRODANOVIĆ, Bibliografija stočarstva i stočnih proizvoda Jugoslavije od 1710–1956. Sarajevo 1956. pp.408. [6506.]

R. TRAUTMANN, Za rozvoj živočišmé výroby. Universita: Knihovna: Výběrový seznam (no.12): Brně 1956. pp.[44]. [78.]*

ANNA KIRILLOVNA ORLOVA, Передовые методы содержания скота. Рекомендательный список литературы. Государственная... публичная библиотека имени М. Е. Салтыкова-Щедрина: Ленинград 1956. pp.23. [75.]

VERA VLADIMIROVNA GNUCHEVA, Выращивание сельскохозяйственных животных и птицы. Рекомендательный указатель литературы. Государственная... публичная библиотека имени М. Е. Салтыкова-Щедрина: Ленинград 1956. pp.20. [50.]

RAMON F[REDERICK] ADAMS, The rampaging herd. A bibliography of books and pamphlets on men and events in the cattle industry. Norman, Okla. [1959]. pp.xix.464. [2651.]

L. M. VADIKOVSKAYA [*and others*], Животноводство. Рекомендательный указатель литературы. Государственная . . . библиотека

СССР имени В. И. Ленина: В помощь организаторам сельскохозяйственного производства: Москва 1959. pp.243. [2250.]

ALMA G. CULEY, Bibliography of beef production in Australia (*c.*1930–1958). McMaster animal health laboratory: Division of animal health: Sydney 1961. pp.viii.239. [4000.]*

VĚRA KLIMEŠOVÁ, Nová technologie živočisné výroby. Universita: Knihovna: Výběrový seznam (no.58): Brně 1961. pp.12. [40.]

ÍNDICE bibliográfico del Centro de investigación y de estudios avanzados. Sección 5ª. . . . II. Agricultura, zootécnia e industrias de la alimentación. Instituto politécnico nacional: México 1962 &c.
*in progress.*

**Lonchocarpus.** [*see also* **Derris.**]

R[URIC] C[REEGAN] ROARK, Lonchocarpus (barbasco, cube, and timbo) — a review of recent literature. Bureau of entomology and plant quarantine: Division of insecticide investigations: Washington 1938. pp.174. [480.]*

INSECT pests of stored roots of derris and cubé (Lonchocarpus). Science library: Bibliographical series (no.649): 1948. single leaf. [15.]*

**Madder.**

J[ULES] CLOÜET and JOS[EPH] DÉPIERRE, Dictionnaire bibliographique de la garance. 1879. pp. xxxi.72. [1200.]

**Maine, U.S.A.**

SAMUEL L. BOARDMAN, Agricultural bibliography of Maine. Augusta 1893. pp.117. [1250.]

**Maize.**

SELECT list of references on the corn products industry. Library of Congress: Washington 1913. ff.2. [20.]*

M. HELEN KEITH, A bibliography of investigations bearing on the composition and nutritive value of corn and corn products. National research council: Washington 1920. pp.xiii.178. [1250.]*

A SHORT list of references on the industrial uses of corn. Library of Congress: [Washington] 1924. ff.5. [42.]*

LOUISE O[LDHAM] BERCAW, ANNIE M[URRAY] HANNAY and NELLIE G[ENEVA] LARSON, Corn in the development of the civilization of the Americas. A selected and annotated bibliography. Compiled . . . under the direction of Mary G. Lacy. Department of agriculture: Bureau of agricultural economics: Agricultural economics bibliography (no.87): Washington 1940. pp.iv.195. [405.]*

ПЕРЕДОВИЙ досвід вирощування високих урожаїв озимої пшениці та кукурудзи в Українскій РСР. Короткий бібліографічий покажчик. Академия наук УРСР: Библиотека: Київ 1953. pp.20. [99.]

E[LENA] V[LADIMIROVNA] PARKHOMENKO and E. M. TSVETAEVA, Гибридная кукуруза. Аннотированный указатель литературы. Центральная научная сельскохозяйственная библиотека [&c.]: Москва 1956. pp.36. [60.]

ANGELINA MARTÍNEZ and C[ATHERINE] NOEL JAMES, Corn. Bibliography of publications available in the institute library. Inter-american institute of agricultural sciences: Orton memorial library: Bibliographical list (no.3): Turrialba, Costa Rica 1960. pp.xv.520.[xiii].521–1130. [10,734.]

I. E. EMELYANOV, *ed.* Кукуруза. Библиографический указатель отечественной литературы за 1794–1959 гг. Всесоюзная . . . академия сельскохозяйственных наук имени В. И. Ленина: Москва 1961. pp.364. [4519.]

**Mango.**

MORTIMER J. SOULE, A bibliography of the mango (mangifera indica L.). Florida mango forum *and* University of Miami: Coral Gables [1950]. pp.vi.89. [1300.]

DISEASES and pests of the mango. Pakistan national scientific & technical documentation centre: Pansdoc bibliography (no.262): Karachi 1960. pp.11. [150.]*

**Manure.**

ABRIDGMENTS of the specifications relating to manure. Patent office: 1857. pp.vi.58. [125.]

*covers the period 1721–1855.*

— 1856–1866. (With . . . indexes to parts I and II.) 1877. pp.xi.57–214. [400.]

*the abridgments for 1867–1876 are included in the volume devoted to drains and sewers, which is entered under Hygiene: Public, above; thereafter no more published.*

**Margarine.**

EFFECTS of fats on animals, with special reference

to the comparative value of butter and margarine (1933–1958). Science library: Bibliographical series (no.762): [1958]. pp.5. [100.]*

**Mate.**

ENRIQUE SPARN, Bibliografía de la yerba mate (*ilex paraguayensis* St.-Hil.). Academia nacional de ciencias: Miscelánea (no.22): Córdoba [Argentina] 1937. pp.20. [200.]

**Meal.** [*see also* **Grain.**]

THE COMPARATIVE feeding values of meat meal, fish meal, and whale meat meal. Science library: Bibliographical series (no.260): 1935 [1936]. ff.5. [52.]*

— [another edition]. . . . (no.508): 1939. ff.5. [70.]*

PROCESSING of fish meals. Science library: Bibliographical series (no.626): 1946. ff.2. [21.]*

**Milk.** [*see also* **Dairy farming.**]

[BARON] HENRI [JAMES NATHANIEL CHARLES] DE ROTHSCHILD, Bibliographia lactaria. Bibliographie générale des travaux parus sur le lait et sur l'allaitement jusqu'en 1899. 1901. pp.xii.584. [9500.]

— — Premier supplément. 1901. pp.vi.98. [1500.]

— — Deuxième supplément. 1902. pp.[v].106. [1578.]

[V. R. Y. CROESEN], Literatur over melk, zuivel en zuivelbereiding geschreven door Nederlanders. Koninklijke nederlandsche landbouwvereeniging: Mededeelingen en berichten (no.8): 's-Gravenhage 1913.

LIST of references on the milk industry. Library of Congress: Washington 1916. ff.27. [395.]*

LIST of references on the by-products of the milk industry. Library of Congress: [Washington] 1917. ff.4. [44.]*

[JEAN MASURE], Répertoire bibliographique sur la question du lait. Bruxelles 1924. ff.[ii].24. [265.]

PAPERS published in the United States on the pasteurization of milk. I. General, plant and technique. Science library: Bibliographical series (no. 351): 1937. ff.9. [177.]*

— II. Nutrition, hygiene, manufacturing quality ... (no.355): 1937. ff.21. [400.]*

MANUALE lactis. Periodisch erscheinendes handbuch der milchwirtschaftlichen weltliteratur. Folge II. Klasse 0 [&c.]. Nürnberg.

0. Milch in der landwirtschaft. Bearbeitet von Klaus Börger. [1952]. pp.xi.220. [2000.]
1. Milch als nahrungsmittel. 1953. pp.xi.154. [1000.]

*in progress.*

**Millet.**

BIBLIOGRAPHY on the breeding and genetics of the millets and sorghums. Imperial bureau of plant genetics: School of agriculture: Cambridge 1932. ff.[ii].21. [100.]*

**Morus.**

[P. A. BOISSIER SAUVAGES DE LA CROIX], Catalogue des auteurs qui ont écrit sur les vers à soie & sur les mûriers. [Nîmcs 1763]. pp.8. [100.]

ROBERTO DI TOCCO, Bibliografia del filugello (bombyx mori L.) e del gelso (morus alba L.). Ente nazionale serico: Milano 1927. pp.xliii.262. [3850.]

**Mycosis.**

H[AROLD] L[E ROY] CHUTE and D[AVID] C[HARLES] O'MEARA, A bibliography avian mycosis. Maine agricultural experiment station: Miscellaneous publication (no.631): [Orono] 1957. pp.81. [521.]

**Nematodes.**

BIBLIOGRAPHY on insect pest resistance in plants (with a supplement on resistance to nematodes). Imperial bureau of plant breeding and genetics: School of agriculture: Cambridge 1944. pp.40. [nematodes: 102.]

**Nutmegs.**

INSECT pests of stored nutmegs and myrobalans. Science library: Bibliographical series (no.647): 1948. single leaf. [12.]*

**Nuts.**

LIST of references on the cultivation, marketing, and uses of nuts. Library of Congress: [Washington] 1924. ff.10. [105.]*

**Oats.**

OAT breeding bibliography. Imperial bureau of

plant genetics: Cambridge 1931. ff.[ii].21. [125.]*

UTILIZATION of oat hulls. Science library: Bibliographical series (no.249): 1936. ff.2. [30.]*

MILLING of oats. Science library: Bibliographic series (no.665): 1948. ff.2. [24.]*

OAT milling. Public library of South Australia: Research service: [Adelaide] 1958. ff.3. [35.]*

JORGE MARTINEZ RÍOS, Bibliografía antropológica y sociológica del estado de Oaxaca. Universidad nacional: Instituto de investigaciones sociales: México 1961. pp.154.

**Oklahoma agricultural experiment station.**

HELEN J. POULTON, Analytical index and list of publications of the Oklahoma agricultural experiment station, 1891–1956. Oklahoma state university: [Stillwater 1957]. pp.[iii].128. [1048.]*

**Olives.**

M[AMIE] I[DELLA] HERB and LOUISE O[LDHAM] BERCAW, Olives and olive oil. Selected references. [Department of agriculture: Washington] 1929. ff.8. [60.]*

GIACINTO DONNO, Bibliografia sistematica dell'olivo e dell'olio di olivo. Ente economico della olivicoltura: Roma 1943. pp.xvi.133. [5000.]

**Peach.**

[FRANÇOIS] THIÉRION, Revue bibliographique des principaux ouvrages français où il est traité de la taille des arbres fruitiers et particulièrement du pêcher. Troyes 1843. pp.108. [100.]

LOUISE O[LDHAM] BERCAW, The peach industry in the United States. A selected list of references on the economic aspects of the industry, including some references relating to Canada. Department of agriculture: Bureau of agricultural economics: Agricultural economics bibliography (no.8): Washington 1925. pp.35. [202.]*

MARK H[UGHLIN] HALLER, Handling, transportation, storage, and marketing of peaches. A digest of recent contributions to the knowledge of physical and biological phases of the subject. Department of agriculture: Bibliographical bulletin (no.21): Washington 1952. pp.i.105. [233.]

**Pear.**

W. H. RAGAN, Nomenclature of the pear. A

catalogue-index of the known varieties referred to in american publications from 1804 to 1907. Department of agriculture: Bureau of plant industry: Bulletin (no.126): Washington 1908. pp.268. [12,500.]

**Phenothiazine.**

BIBLIOGRAPHY of phenothiazine as an anthelminthic. Imperial bureau of agricultural parasitology: St. Albans 1942. pp.[ii].6. [120.]

**Phosphorus.**

E[RNEST] B[ROWNING] FORBES and M. HELEN KEITH, A review of the literature of phosphorus compounds in animal metabolism. Ohio agricultural experiment station: Technical series: Bulletin (no.5): Wooster 1914. pp.748. [2500.]

**Phylloxera.**

PHYLLOXERA . . . 1950–1959. Science library: Bibliographical series (no.777): [1960]. pp.7. [150.]*

**Phytophaga.**

J. S. WADE, An annotated bibliography of the

Hessian fly, phytophaga destructor (Say). Department of agriculture: Miscellaneous publication (no.198): Washington 1934. pp.100. [1256.]

**Pig.**

ROSLYN KUPPERMAN RUBINS, KATHRYN L. BARINGER and DOROTHY B. SKAU, Improved cottonseed meals for feeding poultry and swine, 1954–55. An annotated bibliography. Department of agriculture: Agricultural research administration: Southern utilization research branch: Southern regional research laboratory: [New Orleans 1957]. pp.91. [159.]*

A BIBLIOGRAPHY of farm buildings research 1945–1958. Part 1. Buildings for pigs. Agricultural research council: 1959. pp.36. [123.]

**Pineapple.**

PINEAPPLES, Science library: Bibliographical series (no.94): [1933]. ff.[i].2. [25.]*

**Plantations.**

EDGAR T[RISTRAM] THOMPSON, The plantation: a bibliography. Pan american union: Department of cultural affairs: Social science section: Social science monographs (no.4): Washington 1957. pp.vii.93. [1347.]*

*also published in spanish.*

**Poppy.**

CULTIVATION of poppies for oil. Science library: Bibliographical series (no.227): 1936. single sheet. [11.]*

**Pork.**

PRESERVATION of fresh pork and bacon. Science library: Bibliographical series (no.268): 1936. ff.2. [33.]*

**Potash.**

LIST of references on potash deposits in Spain and Chile. Library of Congress: Washington 1916. ff.2. [21.]*

LIST of references on the potash industry. Library of Congress: [Washington] 1918. ff.13. [162.]*

J[ULIUS] F[REDERICK] T[HOMAS] BERLINER, Potash bibliography to 1928 (annotated). Review . . . of technical literature on potash salts (including the alunites) and their foreign occurrences. Bureau of mines: Bulletin (no.327): Washington 1930. pp. ii.578. [3967.]

BIBLIOGRAPHY of literature on potash as a plant nutrient. American potash institute: Washington.*

i. July–September 1939. [By] Dorothy H. Jameson and Catherine M. Schmidt. 1940. pp.[iii].80.xvii. [322.]
ii. October 1939–March 1940. pp.[iv].108. xxv. [505.]
iii. April–August 1940. By C. M. Schmidt, D. H. Jameson, Virginia G. Specht. 1946. pp.[iii].91.xxix. [519.]
iv. September–December 1940. pp.[iv].82. xxv. [275.]

v. January–March 1941. pp.[ii].55.xviii [264.].
vi. April–June 1941. pp.[iv].46.xvi. [260.]
vii. July–September 1941. Compiled by Helen N. Hudgins [*and others*]. 1952. ff.99. [323.]

viii. October–December 1941. 1953. pp.100. [312.]
ix. January–March 1942. 1953. pp.57. [190.]
x. April–June 1942. 1953. pp.64. [229.]

*in progress?*

**Potatoes.**

THE MANURING of potatoes, 1930–7. Science library: Bibliographical series (no.388): 1938. ff.16. [320.]*

SELECT list of references on the physiology of the potato, with special reference to dormancy. Science library: Bibliographical series (no.504): 1939. ff.5. [70.]*

EXTRACTION of starch from potatoes. Science library: Bibliographical series (no.518): 1940. ff.2. [40.]*

THE DRYING of potatoes. Science library: Bibliographical series (no.519): 1940. ff.4. [80.]*

DEAN H[UMBOLDT] ROSE and HAROLD T[HURSTON] COOK, Handling, storage, transportation, and utilization of potatoes. A digest of information . . . published mostly from 1938 to 1948. Department of agriculture: Bibliographical bulletin (no.11): Washington 1949. pp.163. [650.]

A BURGONYATERMESZTÉS magyar szakirodalma 1774–1959. (Bibliográfia). Országos mezőgazdasági könyvtár és dokumentaciós központ: Budapest 1960. pp.[iv].200. [4918.]*

A BIBLIOGRAPHY of farm buildings research 1945–1958. Part 2. Buildings for potato storage. Agricultural research council: 1960. pp.31. [90.]

**Potato-beetle.**

A. ČERMÁKOVÁ [*and others*], Mandelinka bramborová (leptinotarsa decemlineata Say). Přehled literatury. Československa akademie zemědelských věd: Prameny literatury (vol.iii, no.4): Praha 1958. pp.60. [1200.]

**Poultry.**

EMMA B. HAWKS, A bibliography of poultry. Department of agriculture: Library bulletin (no. 18): Washington 1897. ff.32. [900.]

CATALOGUE of books on gardening, poultry & bees. Municipal libraries: Leicester 1927. pp.15. [poultry: 27.]

— — [another edition]. 1933. pp.15. [19.]

R[EGINALD] C[RUNDALL] PUNNETT, Notes on old poultry books. With a bibliography up to 1880 by E. Comyns Lewer and R. C. Punnett. 1930. pp.40. [125.]

*see also Jeffers, 1954, below.*

LOUISE O[LDHAM] BERCAW, The poultry industry. A selected list of references on the economic aspects of the industry, 1920–1927. Department of agriculture: Bureau of agricultural economics: Agricultural economic bibliography (no.24): Washington 1928. pp.[ii].iv.106. [341.]*

LIST of books on poultry farming (including fowls, ducks, geese, pheasants and pigeons) published during the period January 1926–June 1936. Science library: Bibliographical series (no.271): 1936. ff.7. [134.]*

ALEXIS L[AWRENCE] ROMANOFF, Popular publications in the field of poultry industry, issued by the agricultural experiment station of the United States, ending June 1941. Cornell university: Agricultural experiment station: Ithaca, N.Y. 1942. ff.111. [1050.]*

HELEN E. HENNEFRUND, Bibliography on the poultry industry in countries other than Canada and the United States. Department of agriculture: Library: Library list (no.12): Washington 1944. pp.113. [982.]*

STILBESTROL and derivatives in poultry production. Annotated bibliography. Merck & co.: Rahway, N.J. 1945. ff.[i].14. [17.]*

LIST of references to the use of poultry feathers. Science library: Bibliographical series (no.718): 1953. ff.2. [24.]*

FREDERICK HARDS JEFFERS, A selection of old poultry books, english and american . . . and a digest of Punnett's "Notes on old poultry books". Windsor, N.Y. 1954. pp.26. [100.]

ROSLYN KUPPERMAN RUBINS, KATHRYN L. BARINGER and DOROTHY B. SKAU, Improved cottonseed meals for feeding poultry and swine, 1954–55. An annotated bibliography. Department of agriculture: Agricultural research administration: Southern utilization research branch: Southern regional research laboratory: [New Orleans 1957]. pp.91. [159.]*

**Rice.**

LIST of references on the rice industry. Library of Congress: Washington 1916. ff.2. [12.]*

— [another edition]. List of references on rice and the rice industry. 1920. ff.14. [146.]*

RICE breeding bibliography. Imperial bureau of plant genetics: Cambridge [1931]. ff.[ii].24. [125.]*

G[RAHAM] V[ERNON] JACKS and M[ARIAN] KATHLEEN MILNE, An annotated bibliography of rice soils and fertilizers. Commonwealth bureau of soil science [Harpenden] [and] Food and Agriculture organization of the United Nations: Rome 1954. pp.v.180. [952.]

LORENZO VANOSSI, Bibliografia del riso fina al 1899. Ente nazionale risi: Ufficio studi: [Milan 1959]. pp.47. [800.]

INTERNATIONAL bibliography of rice research. International rice research institute: New York 1963. pp.lvi.881. [7274.]*

**Rural life and problems.**

1. *Bibliographies*

COUNTRY books. A list of bibliographies. National book league *and* National federation of young farmers' clubs. [1953]. pp.12. [26.]

2. *General*

BARNICOTT & SON, The county gentleman's reference catalogue. Taunton 1883. pp.126. [1750.]

[V. A. N. SAWYER], A list of some recent works on housing and on rural problems. Edited by Leslie A. St. L. Toke. Catholic truth society: [1909]. pp.32. [299.]

SELECT list of references on rural life. Library of Congress: [Washington] 1911. ff.7. [46.]*

— Additional references. 1915. ff.3. [32.]*

LIST of references on city and country life. Library of Congress: [Washington] 1921. ff.8. [103.]*

LIST of references on rural life and culture. Bureau of education: Library leaflet (no.16): Washington 1922. pp.10. [150.]

— [second edition] . . . (no.26): 1924. pp.12. [200.]

MARY G[OODWIN] LACY, A beginning of a bibliography of the literature of rural life. Department of agriculture: Bureau of agricultural economics: Agricultural economics bibliography (no.3): Washington 1924. pp.[ii].20. [150.]*

A LIST of books dealing with social, economic and religious aspects of country life. Library of Congress: Washington 1924. ff.6. [48.]*

MARY G[OODWIN] LACY, *ed.* Research in rural economics and rural sociology in the southern states since 1920. A list of the published, unpublished, and current studies. Department of agriculture: Bureau of agricultural economics: Agricultural economics bibliography (no.10): Washington 1926. pp.iv.44. [407.]

LOAN books for rural workers. University of Nanking: College of agriculture and forestry: Bulletin (no.16): [Nanking] 1926. pp.11.

RURAL life from the aspect of the social worker. Russell Sage foundation: Library: Bulletin (no.98): New York 1929. pp.4. [75.]

LOUISE O[LDHAM] BERCAW, Rural standards of living. A selected bibliography. Department of agriculture: Bureau of agricultural economics: Agricultural economics bibliography (no.32): Washington 1930. pp.[ii].ii.124. [282.]*

LOUISE O[LDHAM] BERCAW, Advantages and disadvantages of country life. Selected references. Department of agriculture: Bureau of agricultural economics: Agricultural economics bibliography (no.37): Washington 1932. pp.[ii].30. [200.]*

I[OEL] N[AFTALEVICH] KOBLENTS, Экономика и организация отдельных отраслей сельского хозяйства. Всесоюзная ассоциация с.-х. библиограии: Ежегодник аграрной литературы СССР за 1928 г. (vol.i, part 3): Москва [1934]. pp.347. [9000.]

PUBLICATIONS relating to farm population and rural life issued at the various state colleges of agriculture. Department of agriculture: Bureau of agricultural economics: Division of farm population and rural welfare: Washington 1936. pp. [ii].27. [400.]*

GÜNTHER FRANZ, Bücherkunde zur geschichte des deutschen bauerntums. Der forschungsdienst (sonderheft 9): Neudamm &c. 1938. pp.97. [1422.]

READERS' guide to rural life. Library association: County libraries section: Readers' guide (no.27): 1939. pp.16. [350.]

MARGARET T[HOMPSON] OLCOTT, Rural psychology. A partial list of references. Department of agriculture: Bureau of agricultural economics: Agricultural economic bibliography (no.78): Washington 1939. pp.vi.76. [183.]*

RETURN to husbandry. An annotated list of books dealing with the history, philosophy and craftsmanship of rural England, and intended to suggest alternatives to commercialization and mechanization. . . . Edited by Edmund [Charles] Blunden. 1943. pp.32. [50.]

LAVERNE BURCHFIELD, Our rural communities. A guidebook to published materials on rural problems. Public administration service: Chicago 1947. pp.xiii.201. [1000.]

ROBERTA C[HAPMAN] WATROUS and JOHN M[ILTON] MCNEILL, Rural community organization. A list of references. Department of agriculture: Library list (no.46): Washington 1949. pp.51. [1104.]

RURAL economics: a selected list of books. Library of Congress: Washington 1949. ff.3. [12.]*

RURAL life. Library association: Readers' guides (no.24): 1954. pp.39. [750.]

[A. D. EIKHENGOLTS *and others*], Общая библиография, учебник для библиотечных институтов. Управление учебных заведений: Москва 1957. pp.464. [2000.]

RURAL sociology. Unesco: Current sociology (vol.vi, no.1): Paris 1957. pp.75. [597.]

HENRI MENDRAS, *ed.* Les sociétés rurales françaises. Éléments de bibliographie. Bibliographies françaises de sciences sociales (no.3): 1962. pp.124. [506.]

**Sawdust.**

SAWDUST as a fertilizer and mulch. Science library: Bibliographical series (no.651): 1948. ff.2. [29.]*

**Seed.** [*see also* **Grain.**]

HYAC[INTHE] LONAY, Analyse coordonnée des travaux relatifs à l'anatomie des téguments séminaux. Université de Liége: Institut botanique: Archives (vol.iv, no.2): Liége 1904. pp.146. [300.]

THE DRY disinfection of seed, 1930–36. Science library: Bibliographical series (no.286): 1935 [1937]. ff.17. [250.]*

— — 1930–1939. . . . (no.499): 1939. ff.19. [275.]*

SEED drills and planters, 1920–37. Science library: Bibliographical series (no.370): 1938. ff.3. [56.]*

**Sesame.**

MAUDE K. SWINGLE, Sesame. A list of references.

Department of agriculture: Library: Library list (no.20): Washington 1945. pp.18. [216.]*

**Sheep.**

EMMA B. HAWKS, Cattle, sheep and goat production in the range country. A selected list of publications issued by the U.S. Department of agriculture and the agricultural colleges, experiment stations and departments of agriculture of the seventeen range states. Department of agriculture: Bibliographical contributions (no.19): Washington 1928. pp.78. [1000.]*

BIBLIOGRAPHY on the biology of the fleece. Imperial bureau of animal genetics: Edinburgh 1931. pp.32. [550.]

DUŠAN PRODANOVIĆ, Bibliografska građa o ovčarstvu i vunarstvu, 1774–1950. Beograd 1950. pp.39. [540.]

EVELYN RUTH KAHN, Karakul sheep in south west Africa. A bibliography. University of Capc Town: School of librarianship: [Capetown] 1959. pp. [ii].iv.22. [200.]

**Silage.**

THE PRODUCTION and utilization of silage. A

review of world literature in abstracts. Commonwealth bureau of pastures and field crops [Hurley, Berks.]: Bulletin (no.40): Penglais [1950]. pp.[v]. 309. [969.]

## Soil.

### 1. *Periodicals*

LISTE des périodiques reçus au Bureau interafricain des sols. 1961. ff.[i].27. [245.]*

### 2. *General*

LIST of publications of the Bureau of soils. Washington 1905. pp.16.

— [another edition]. 1917. pp.21.

SOILS of interest to California citrus growers. Public library: Bulletin (no.111): Riverside, Cal. 1914. pp.16. [200.]

CORA L. FELDKAMP and CATHERINE E. PENNINGTON, List of the publications on soils issued by the state agricultural experiment stations of the United States through 1926. Department of agriculture: Bibliographical contributions (no.15): Washington 1927. pp.81. [1250.]*

EMMA B[EATRICE] HAWKS and CHARLOTTE TROLINGER, List of the publications on soils by the U.S. Department of agriculture, 1844–1926. Department of agriculture: Bibliographical contributions (no.14): Washington 1927. pp.63. [750.]*

B. G. POLYNOV [POLUINOV], Contributions of russian scientists to paleopedology. Academy of sciences of the Union of the Soviet Socialist Republics: Russian pedological investigations (no.viii): Leningrad 1927. pp.33. [96.]

LITERATURSAMMLUNG aus dem gesamtgebiet der agrikulturchemie. Technische hochschule: Agrikulturchemisches institut: München.

i. Bodenkunde. Von H[ans] Niklas, F. Czibulka und A. Hock. 1931. pp.xxxvi. 1008. [25,000.]
ii. Bodenuntersuchung. 1931. pp.xxviii.199. [4000.]

PUBLICATIONS relating to soils and fertilizers. List no.1[–77]. January–April 1931[–December 1937]. Imperial bureau of soil science: Rothamsted experimental station: Harpenden 1931–1938.*

*revised and continued as:*

Bibliography of soil science, fertilizers and general agronomy. Imperial [Commonwealth]

bureau of soil science: Harpenden.
1931–1934. pp.v–xxxi.475. [6000.]
1934–1937. pp.v–xv.540. [8000.]
1937–1940. pp.iii–xxiii.504. [5500.]
1940–1944. pp.xxv.567. [5500.]
1944–1947. pp.xxvii.451. [5000.]
1947–1950. pp.xxviii.535. [5500.]
1950–1953. pp.xxviii.711. [7000.]

PUBLICATIONS on soil science issued from the empire overseas. Imperial bureau of soil science: Rothamsted experimental station: Harpenden.
1931. ff.[ii].15. [300.]
1932. ff.[ii].24. [500.]
1933. ff.[ii].22. [400.]
1934. ff.[ii].14. [200.]
[*continued as:*]
Soil research in the British empire.
1935. ff.[ii].22. [100.]
*no more published.*

PAMPHLETS and reprints obtainable on loan. Imperial bureau of soil science: Harpenden 1932. ff.[ii].59. [700.]*

THE SOILS of Denmark. Science library: Bibliographical series (no.26): [1932]. ff.[i].2. [19.]*

BIBLIOGRAPHY on waterlogging of soils. Science

library: Bibliographical series (no.39): [1932]. ff.[i].1. [16.]*

LILLIAN H. WIELAND, Bibliography on soil conservation. . . . Revised by June Henderson. Department of agriculture: Soil conservation service: [Washington 1936]. pp.[ii].iv.179. [2000.]*

STANLEY H. GAINES, Bibliography on soil erosion and soil and water conservation. . . . With abstracts by Francesca Vincent, Marion Bloom and James F. Carter. Department of agriculture: Miscellaneous publication (no.312): Washington 1938. pp.v.651. [4388.]

C[HARLES] R[ANGER] ENLOW, Review and discussion of literature pertinent to crop rotation for erodible soils. Department of agriculture: Circular (no.559): Washington [1940]. pp.51. [158.]

SOIL stabilisation, especially with bituminous materials. Science library: Bibliographical series (no.565): 1942. ff.2. [30.]*

JUAN DE DIOS TEJADA Y SAÍNZ, Bibliografía de los suelos de Cuba. Santa María del Rosario 1943. pp.viii.77. [200.]

MARGARET LUSCOMBE, Soil. A bibliography. University of Cape Town: School of Librarianship: Bibliographical series: [Capetown] 1945. ff.55. [410.]*

ĐORĐE ZLOKOVIĆ, Библиографска грађа за педолошко проучавање југословенских земаља. Београд 1949. pp.139. [860.]

RUBÉN H. MOLFINO, Reseña histórica de los trabajos sobre suelos, realizados por el Ministerio de agricultura, desde su fundación. Instituto de suelos y agrotecnica: Buenos Aires 1952. pp.12. [50.]

FLORENCE [EVELYN] COZZI, Soil mechanics photoelastic stress analysis. An abstract bibliography. QM research and development laboratories: Technical library: Bibliographic series (no.27): Philadelphia [1953]. pp.16. [62.]

SCHRIFTTUM über bodenmechanik (bautechnische bodenkunde). Forschungsarbeiten an dem strassenwesen (n.s., vol.10 &c.): Bielefeld.

[i]. Bearbeitet von . . . H[ans] Petermann. . . . (n.s. vol.10): 1953. pp.xvi.526. [8989.]
ii. . . . (n.s. vol.32): 1957. pp.xxvii.254. [4128.]
iii. . . . (n.s. vol.46): 1961. pp.xxvii.363. [5479.]

*in progress.*

HAROLD T. PINKETT, Preliminary inventory of the records of the Bureau of plant industry, soils, and agricultural engineering. National archives: Preliminary inventories (no.66): Washington 1954. pp.v.49. [very large number.]*

EUGENE Y. HUANG [YU-CHING HUANG], Selected bibliography on soil stabilization. University of Michigan: Engineering research institute: Ann Arbor 1954. ff.[i].154. [554.]*

PUBLICATIONS on soils. Imperial college of tropical agriculture: St Augustine, Trinidad 1955. pp.[5]. [100.]*

ANNOTATED bibliography of geologic and soils literature of western north Pacific islands. United States army forces far east: Office of the engineer: [*s.l.*] 1956. pp.[iii].884. [7500.]*

SOIL compaction and stabilization. A selected list of titles. Department of the army: Army library: Special bibliography (no.6): [Washington] 1956. pp.[ii].42. [250.]*

SOIL inoculation. Public library of South Australia: Research service: [Adelaide] 1956. ff.5. [58.]

G[EORGES] PLAISANCE, Bibliographie des sols de

France. Centre d'études et de documentation paléontologiques: Annales (no.21): Dole 1957. pp.245. [3000.]*

G[EORGES] PLAISANCE, Bibliographie des sols forestiers. (Répertoire international). Centre d'études et de documentation paléontologiques: Annales (no.25): 1957. pp.165. [2250.]*

CONTRIBUTIONS by norwegian authors to the field of soil mechanics and foundation engineering. Norges geotekniske institutt: Oslo 1957 &c.*
*in progress.*

ISTVÁN TERTS, A magyar talajtani irodalom bibliográfiája, 1914–1953. Budapest 1957. pp.88. [1418.]

FLOYD O[WEN] SLATE and A. W. JOHNSON, Stabilization of soil with calcium chloride. National research council: Highway research board: Bibliography (no.24): Washington 1958. pp.vii.96. [449.]*

E[LIZAVETA] K[HRISTIANOVNA] BUICHKOVA and YU[LIYA]. O[SIPOVNA] BIBILO, Почвы Юго-Востока европейской части СССР. Библиографический указатель. Университет: Научная библиотека: Библиография Саратов-

xxvii. 1887. pp.x.354. [150.]
xxviii. 1888. pp.xiv.377. [150.]
xxix. 1889. pp.xii.336. [150.]
xxx. 1890. pp.x.338. [100.]
xxxi. 1891. pp.xii.372. [200.]
xxxii. 1892. pp.x.400. [200.]
xxxiii. 1893. Herausgegeben von Joh. Bock. pp.xii.356. [250.]
xxxiv. 1894. pp.x.354. [250.]
xxxv. 1895. pp.xi.379. [300.]
xxxvi. 1896. pp.xi.487. [350.]
xxxvii. 1897. pp.xi.378. [350.]
xxxviii. 1898. pp.xi.326. [400.]
xxxix. 1899. pp.xi.338. [500.]
xl. 1900. pp.xi.533. [450.]
xli. 1901. pp.xii.374. [750.]
xlii. 1902. pp.xiii.371. [650.]
xliii. 1903. pp.xi.313. [750.]
xliv. 1904. pp.x.307. [750.]
xlv. 1905. pp.x.321. [750.]
xlvi. 1906. pp.xi.351. [850.]
xlvii. 1907. pp.x.342. [850.]
xlviii. 1908. pp.xii.287. [750.]
xlix. 1909. pp.xii.361. [850.]
l. 1910. pp.xi.319. [750.]
li. 1911. pp.x.278. [750.]

*no more published; vol.xii includes an index to*

*vols.i–xii, vol.xxvi to vols.xiii–xxvi, and vol.xl to vols.i–xl.*

ABRIDGMENTS of specifications relating to sugar, A.D. 1663–1866. Commissioners of patents for inventions: 1871. pp.xix.442. [500.]
*page 416* is inserted between pages 416 and [417].*

ABRIDGMENTS of specifications. Class 127. Sugar. Patent office.
1855–1866. pp.vii.40. [200.]
1867–1876. pp.viii.68. [325.]
1877–1883. pp.viii.60. [300.]
1884–1888. pp.vii.38. [175.]
1889–1892. pp.vii.26. [125.]
1893–1896. pp.viii.20. [100.]
1897–1900. pp.vii.28. [125.]
1901–1904. pp.viii.26. [125.]
1905–1908. pp.vi.22. [100.]
1909–1915. pp.vi.25. [100.]
1916–1920. pp.v.14. [60.]
1921–1925. pp.vi.27. [125.]
1926–1930. pp.vi.34. [150.]
*no more published.*

VERZEICHNISS sämmtlicher schriften über zuckerfabrikation welche von 1865–1881 im deutschen buchhandel erschienen sind. Gracklauer's fachkatalog (no.33): Leipzig 1881. pp.6. [50.]

H[ENRY] LING ROTH, A guide to the literature of sugar. 1890. pp.xv.159. [1400.]

CLARIBEL R[UTH] BARNETT, References to the literature of the sugar beet, exclusive of works in foreign languages. Department of agriculture: Library bulletin (no.16): [Washington] 1897. pp.9. [400.]

LIST of references on the regulation of sugar exportation in foreign countries. Library of Congress: Washington 1904. ff.8. [47.]*

LIST of references on maple sugar as found in the literature of american history, description, travel, etc. Library of Congress: Washington 1906. ff.5. [50.]*

G. W. KIRKALDY, A bibliography of sugar-cane entomology. Hawaiian sugar planters' association: Experiment station: Division of entomology: Bulletin (no.8): Honolulu 1909. pp.73. [300.]

HERMANN HENRY BERNARD MEYER, Select list of references on sugar, chiefly in its economic aspects. Library of Congress: Washington 1910.

pp.238. [2000.]

[—] — Supplementary. 1917. ff.27. [388.]*

A. J. SCULIER, Catalogue des ouvrages spéciaux de langue française concernant le sucre et l'industrie sucrière composant la bibliothèque de m. Lewis S[harpe] Ware. 1911. pp.[iii].340. [3250.]

PATENTS for inventions. Fifty years subject index, 1861–1910. Class 127. Sugar. Patent office: 1913. pp.6. [600.]

LIST of references on cane sugar in early american and local history. Library of Congress: Washington 1913. ff.6. [42.]*

SELECT list of additional references on sugar (supplementing the printed list, 1910). Library of Congress: Washington 1913. ff.16. [199.]*

— [supplement]. 1917. ff.25. [388.]*

SELECTED list of books on sugar manufacture. Public library of New South Wales: Sydney 1919. pp.8. [50.]

LIST of references on maltose, or malt sugar. Library of Congress: [Washington] 1919. ff.4. [32.]*

A LIST of recent documentary publications on sugar issued by the United States government. Library of Congress: [Washington] 1924. ff.8. [71.]*

— [supplement]. Sugar and the sugar industry. 1934. ff.12. [102.]*

ZYGMUNT PRZYREMBEL, Polska bibljografja cukrownicza 1799–1929. Warszawa 1930. pp. [viii].91. [500.]

CLARENCE R[ICHARD] BITTING, Bibliography on sugar. [*s.l.* 1937]. pp.[72]. [1000.]

THE APPLICATION of physico-chemical methods of control to the sugar industry, 1933 to date. Science library: Bibliographical series (no.366): 1938. ff.9. [160.]*

**Tea.**

CORNELIS ADRIAAN, Catalogus auctorum qui de thea scripserunt. Trajecti ad Rhenum 1825. pp. [iv].50. [275.]

LIST of references on the production of tea in India, China, Ceylon, Japan, Java. Library of Congress: Washington 1917. ff.5. [59.]*

LIBRARY catalogue of the Tea bureau. [1951]. pp.25. [250.]

**Texas fever.**

SELECT list of references on Texas fever in cattle. Library of Congress: Washington 1913. ff.13. [113.]*

**Ticks.**

LISTE bibliographique des publications de l'Institut Pasteur d'Algérie relatives aux maladies animales transmises par les tiques. Institut Pasteur d'Algérie: Alger 1956. pp.15. [200.]

**Tobacco.**

LIST of references on tobacco leaf curing. Library of Congress: Washington 1915. ff.2. [16.]*

KATHARINE F[LORENCE] WILLIAMS, Cooperative marketing of tobacco. A selected list of references. Department of agriculture: Bureau of agricultural economics: Agricultural economic bibliography (no.13): Washington 1926. pp.[iv]. 5. [30.]*

BIBLIOGRAPHY of publications relating to tobacco soils and fertilisers. Imperial bureau of soil science: Harpenden 1933. ff.[ii].52. [750.]*

TOBACCO breeding bibliography. Imperial bureau of plant genetics: Cambridge 1937. ff.[ii]. 36. [250.]*

JOSÉ LLORCA, Bibliografía sobre fitotecnica del tabaco. Ministerio de agricultura: Instituto nacional del tabaco: [Buenos Aires 1942]. ff.34. [681.]*

JOSÉ LLORCA, Ensayo de una bibliografía argentina del tabaco. Ministerio de agricultura: Instituto nacional del tabaco: [Buenos Aires 1943]. ff.[84].*

— — [another edition]. Ministerio de agricultura: Dirección general de agricultura: Dirección de tabaco: 1945. pp.79. [563.]

VITTORIO CAMERANI and ANDREA FERNANDEZ, Saggio di bibliografia del tabacco. Europa e bacino del Mediterraneo. Istituto scientifico sperimentale per i tabacchi: Roma 1952. pp.lxiii.204. [2847.]

TOBACCO abstracts. North Carolina agricultural experiment station: Tobacco literature service: Raleigh.*

i. 1957. pp.561. [1798.]
ii. 1958. pp.650. [2194.]
iii. 1959. pp.641. [2107.]
iv. 1960. pp.762. [2308.]
v. 1961. pp. . [2788.]
vi. 1962. pp.798. [2510.]
vii. 1963. pp. . [2965.]
*in progress.*

**Tomato.**

THE EXTRACTION of tomato juice. Science library: Bibliografical series (no.392): 1938. f.1. [12.]*

TOMATO spotted wilt virus — with some references to the history of the tomato. Public library of South Australia: Research service: [Adelaide] 1956. ff.8. [90.]*

**Tung.**

K['AI] HO and H[U] LIU, Bibliography on tung tree and tung oil. Government testing bureau: Hankow 1937. pp.vii.175. [1400.]

CULTIVATION of tung oil trees, aleurites spp. Science library: Bibliographical series (no.324): 1937. ff.9. [172.]*

**Vanilla.**

CORA L. FELDKAMP, Vanilla: culture, processing and economics. A list of references. Department of agriculture: Library: Library list (no.13): Washington 1945. pp.28. [362.]*

**Vegetables.**

KATHARINE G. RICE, Bibliography on the preservation of fruit and vegetables in transit and storage, with annotations. Department of agriculture: Library: Bibliographical contributions (no.4): Washington 1922. pp.[iii].76. [750.]*

THE DRYING of fruits and vegetables. References covering the period 1930–1937. Science library: Bibliographical series (no.380): 1938. ff.7. [115.]*

— 1930–1942 . . . (no.587): 1943 ff.10. [200.]*

GARDEN and allotment food. National book council: Book list (no.191): 1942. pp.[4]. [140.]

VEGETABLE and fruit growing. A small selection of the recently published books. Public libraries: Manchester 1944. pp.[6] (folder).

FRUIT juices and vegetables, frozen. Public

library of South Australia: Research service: [Adelaide] 1955. ff.12. [269.]*

MARIE A. JONES, The brining and pickling of cucumbers and other vegetables. A list of publications with abstracts, September 1938–December 1955, on cooperative investigations by U.S. Food fermentation laboratory and North Carolina agricultural experiment station. Department of agriculture: Agricultural research service: Southern utilization research branch: [Raleigh] 1956. pp.[ii].iii.18. [58.]*

F. S. GINZBURG and E. M. TSVETAEVA, Увеличим производство картофеля и овощей. Аннотированный указатель литературы. Центральная научная сельскохозяйственная библиотека: Москва 1956. pp.40. [150.]

## Veterinary science.

[*writings on this subject are often included in early bibliographies of agriculture.*]

### 1. *Periodicals*

VERZEICHNIS der laufenden zeitschriften der landwirtschaftlichen und tierärztlichen bibliotheken in Berlin. Berlin 1936. pp.iii.29. [1103.]*

RÉPERTOIRE des périodiques et des comptes rendus de congrès en compte à la bibliothèque. École nationale vétérinaire: Alfort [1952]. ff.[i].36. [200.]*

2. *History*

CATALOGUE of the historical collection. Books published before 1850. Royal college of veterinary surgeons: Library: 1953. pp.[iii].36. [600.]

3. *General*

FR[ANZ] ER[NST] BRÜCKMANN, Bibliotheca animalis oder verzeichniss der meisten schriften so von thieren und deren theilen handeln. Wolfenbüttel 1743. pp.277.[lvi]. [2000.]

— — Continvatio. 1747. pp.178.[xxx]. [1500.]

JOHANN KARL GOTTLIEB HENZEN, Entwurf eines verzeichnisses veterinarischer schriften und einzelner abhandlungen die zur theoretischen und praktischen kenntniss von pferden, eseln, maulieseln, rindviehe . . . nützen können. Stendal 1781. pp.246. [1650.]

[BERNARDO RODRIGUEZ], Catálogo de algunos autores españoles que han escrito de veterinaria, de equitación, y de agricultura. Madrid 1790. pp.87. [400.]

J[EREMIAS] D[AVID] REUSS, Repertorium commentationum a Societatibus litterariis editarum. ... Tom.XVI. Scientia et ars medica et chirurgica. P. II. Ars veterinaria. Gottingae 1821. pp.[ii].iv. 87. [1000.]

THEODOR CHRISTIAN FRIEDRICH ENSLIN, Bibliotheca veterinaria, oder verzeichniss aller brauchbaren . . . in Deutschland erschienenen bücher über alle theile der thierheilkunde. Berlin &c. 1825. pp.[ii].34. [750.]

— — Zweite . . . auflage. . . . Umgearbeitet von Wilhelm Engelmann. Leipzig 1843. pp.[iii].76. [1250.]

F. S. LEUCKART, Einleitung in die organiatrik und insbesondere die zoïatrik oder thierarzneikunde. Heidelberg 1832. pp.iii–xiv.62. [300.]

JAHRESBERICHT über die fortschritte der thierheilkunde. Erlangen [1851–1865: Würzburg].

1842–1843. [By F. A. Kuers]. pp.[ii].88. [400.]
1844. [By Christian Joseph Fuchs]. pp.60. [300.]
1845. pp.53. [250.]
1846. [By E. Hering]. pp.72. [400.]
1847. pp.68. [300.]
1848. pp.[vii].64. [300.]

[*continued as:*]

C. Canstatt's jahresbericht [&c.].
1849. pp.[iv].83. [400.]
1850. pp.[iii].72. [300.]
[*continued as:*]

Thierheilkunde.
1851. pp.[ii].65. [300.]
1852. pp.[ii].72. [300.]
1853. pp.[ii].68. [300.]
1854. pp.[ii].71. [300.]
1855. pp.[ii].66. [300.]
1856. pp.[ii].70. [300.]
1857. pp.[ii].70. [300.]
1858. pp.[ii].75. [300.]
1859. pp.[ii].62. [300.]
1860. pp.[ii].59. [300.]
1861. pp.[ii].70. [300.]
1862. pp.[ii].69. [300.]
1863. pp.[ii].64. [300.]
1864. pp.[ii].64. [300.]
1865. pp.[iv].68. [300.]
*forms vol.iv of the* Jahresbericht über die fortschritte der gesammten medicin *for the years shown.*

RAMÓN LLORENTE LÁZARO, Compendio de la bibliografía de la veterinaria española. Madrid &c. 1856. pp.viii.204. [250.]

G. W. SCHRADER and EDWARD HERING, Biographisch-literarisches lexicon der thierärzte aller zeiten und länder. Stuttgart 1863. pp.[ii].x.490. [2000.]

ADOLPH BÜCHTING, Bibliotheca veterinaria, oder verzeichniss der seit . . . 1842 bis . . . 1866 im deutschen buchhandel erschienenen bücher und zeitschriften. Nordhausen 1867. pp.68. [700.]

ABRIDGMENTS of specifications relating to farriery; including the medical and surgical treatment of animals 1719–1866. Commissioners of patents for inventions: 1872. pp.xii.184. [200.]
*also contains a bibliography.*
— 1867–1876. 1880. pp.xi.184. [400.]
*no more published.*

JAHRESBERICHT über die leistungen auf dem gebiete der veterinär-medicin. Edited by [W.] Ellenberger . . . und [W.] Schütz [vols.xvii–xxi: and H(ermann) Baum]. Berlin.

i. 1881. pp.[iv].88. [1250.]
ii. 1882. pp.[iv].148. [1750.]
iii. 1883. pp.[iv].176. [1750.]
iv. 1884. pp.[iii].200. [1750.]
v. 1885. pp.iv.174. [1750.]
vi. 1886. pp.iv.218. [2000.]
vii. 1887. pp.iv.160. [1750.]

viii. 1888. pp.iv.229. [2250.]
ix. 1889. pp.iv.222. [2250.]
x. 1890. pp.iv.214. [2250.]
xi. 1891. pp.iv.220. [2500.]
xii. 1892. pp.iv.209. [2500.]
xiii. 1893. pp.iv.222. [2500.]
xiv. 1894. pp.iv.243. [2750.]
xv. 1895. pp.iv.229. [2750.]
— General-register . . . 1881–1895 . . . von Hermann Baum. 1897. pp.[iv].95.
xvi. 1896. pp.iv.225. [2750.]
xvii. 1897. pp.iv.254. [2750.]
xviii. 1898. pp.iv.240. [2750.]
xix. 1899. pp.iv.295. [3000.]
xx. 1900. pp.iv.267. [3000.]
xxi. 1901. pp.iv.312. [3500.]
xxii. 1902. pp.iv.278. [3250.]
xxiii. 1903. pp.iv.316. [3500.]
xxiv. 1904. pp.iv.354. [4000.]
xxv. 1905. pp.iv.436. [4500.]
xxvi. 1906. pp.viii.434. [4500.]
xxvii. 1907. pp.iv.416. [4500.]
xxviii. 1908. pp.iv.460. [4500.]
xxix. 1909. pp.vi.436. [4500.]
xxx. 1910. pp.iv.492. [5000.]
xxxi. 1911. pp.iv.416. [5500.]
xxxii. 1912. pp.v.426. [5500.]

xxxiii. 1913. pp.iv.424. [5500.]
xxxiv. 1914. pp.vii.298. [3500.]
xxxv. 1915. pp.vii.234. [2500.]
xxxvi. 1916. pp.vii.256. [3000.]
xxxvii. 1917. pp.xii.246. [3000.]
[*continued as:*]

Ellenberger-Schütz' jahresbericht [&c.]. Edited by W. Ellenberger and O. Zietzschmann [vols. xxxix–xlvi: and K. Neumann].

xxxviii. 1918. pp.vii.214. [3000.]
xxxix–xl. 1919–1920. pp.v.350. [5000.]
xli–xlii. 1921–1922. pp.v.496. [8000.]
xliii. 1923. pp.v.346. [5500.]
xliv. 1924. pp.v.400. [6000.]
xlv. 1925. pp.vi.404. [6000.]
xlvi. 1926. pp.xi.756+viii.757–1649. [10,000.]

VERZEICHNISS der gesammten literatur über veterinärwissenschaft und populäre thierheilkunde, welche von 1866–1883, im deutschen buchhandel erschienen ist. Gracklauer's fachkatalog (no.41): Leipzig 1883. pp.iv.69. [500.]

STANSŁAW KRÓLIKOWSKI, Bibliografia polska weterynaryi i hodowli zwierząt. Lwów 1891. pp.[vii].167.[vii]. [2060.]

N[IKOLAY] P[ETROVICH] PETROV, Библіографическій указатель русскихъ диссертацій по медицинѣ и ветеринаріи . . . съ 1860 по 1-е іюня 1892 г. включительно. 2 . . . изд. С.-Петербургъ 1892. pp.[ii].101. [2287.]

— — Дополненіе. 1896. pp.[ii].42. [1025.]

— — Второе дополненіе. 1898. pp.32. [617.]

DIE LITTERATUR der veterinär-wissenschaft und deren hilfswissenschaften von 1889–1. Juli 1894. Berlin 1894. pp.34. [300.]

ROBERT KLEE, Bibliotheca veterinaria, oder verzeichnis sämtlicher bis zur gegenwart im deutschen buchhandel erschienenen bücher und zeitschriften. Leipzig 1901. pp.[iii].248. [4500.]

GYULA KÓSSA, Magyar állatorvosi könyvészet, 1472–1904. Magyar országos állatorvos-egyesület: Allatorvosi kézi könyvtár (vol.viii): Budapest 1904. pp.xii.346. [1236.]

TROPICAL veterinary bulletin. Tropical dieseases bureau [*afterwards:* Bureau of hygiene and tropical diseases].

i. 1912–1913. Editor . . . A. Leslie Sheather. pp.v.336. [616.]

ii. 1914. pp.v.213. [383.]

iii. 1915. pp.v.173. [290.]

iv. 1916. Editors . . . A. L. Sheather . . . J. T. Edwards. pp.v.215. [368.]

v. 1917. Editor . . . J. T. Edwards. pp.iv.248. [400.]
vi. 1918. pp.iv.303. [600.]
vii. 1919. pp.iii.319. [600.]
viii. 1920. pp.iii.383. [750.]
ix. 1921. pp.iii.319. [600.]
x. 1922. Editor . . . A. L. Sheather. pp.iii.161. [600.]
xi. 1923. pp.iii.168. [600.]
xii. 1924. pp.iii.161. [600.]
xiii. 1925. pp.iii.165. [600.]
xiv. 1926. pp.iii.177. [600.]
xv. 1927. pp.iii.182. [750.]
xvi. 1928. pp.iii.160. [600.]
xvii. 1929. pp.iii.159. [600.]
xviii. 1930. pp.iii.153. [600.]

*no more published.*

THE REVIEW of applied entomology. Series B: medical and veterinary. Imperial bureau [*afterwards:* institute] of entomology: 1913 &c.

*in progress; details of this work are entered above under Entomology: Medical and veterinary.*

CATALOGUS van de bibliotheek der Veeartsenijkundige faculteit. Rijksuniversiteit: Utrecht 1926–1928. pp.[viii].445+[x].449–808+[v].809–973. [9000.]

— Derde supplement. 1930. pp.[iii].76. [600.] *the first and second supplements form part of the main work; a fourth supplement was issued in (1931) in the form of broadsides.*

EXHIBITION at British museum, Bloomsbury, of manuscripts and early english printed books dealing with veterinary medicine. Eleventh international veterinary congress: 1930. pp.[6]. [35.]

INDEX veterinarius. Imperial [Commonwealth] bureau of animal health: Weybridge.

i. 1933. ff.[i].xxxvi.304 + [i].iv.354 + [i].v.532+[i].iii.[478]. [15,000.]*
ii. 1934. ff.[i].vi.[383]+[i].ii.[295]+[i].iii.[396]+[i].ii.[343]. [12,500.]*
iii. 1935. ff.[i].iv.[408]+[i].[358]+[ii].[474]+[ii].437. [15,000.]*
iv. 1936. pp.[viii].137+[v].126+[v].122+[v].124. [15,000.]
v. 1937. pp.[iv].129+viii.106+viii.133+vii.100. [15,000.]
vi. 1938. pp.viii.255+viii.216. [15,000.]
vii. 1939. pp.viii.249+viii.236. [15,000.]
viii. 1940. pp.viii.167+viii.170. [10,000.]
ix. 1941. pp.viii.132+viii.166. [8500.]
x. 1942. pp.viii.180+viii.194. [12,500.]
xi. 1943. pp.viii.174+viii.185. [12,500.]

xii. 1944. pp.viii.201+viii.247. [15,000.]
xiii. 1945. pp.viii.291. [12,500.]
xiv. 1946. pp.viii.368. [17,500.]
xv. 1947. pp.viii.431. [20,000.]
xvi. 1948. pp.viii.431. [20,000.]
xvii. 1949. pp.viii.431. [20,000.]
xviii. 1950. pp.viii.84+viii.285. [17,500.]
xix. 1951. pp.viii.68+viii.96+viii.94+viii.166. [20,000.]
xx. 1952. pp.xvi.144+xvii.139+xvii.178+xvii.127. [30,000.]
xxi. 1953. pp.xvii.149+ix.123+ix.245+ix.196. [35,000.]
xxii. 1954. pp.xxi.215+ix.160+ix.198+ix.199. [17,500.]
xxiii. 1955. pp. +ix.217+x.192+viii.197. [17,500.]
xxiv. 1956. pp.viii.151+viii.159+viii.160+viii.170. [15,000.]
xxv. 1957. pp.viii.165+viii.165+viii.167+viii.159. [15,000.]
xxvi. 1958. pp.viii.166+viii.167+viii.177+viii.175. [15,000.]
xxvii. 1959. pp.viii.175+viii.183+viii.183+viii.168. [15,000.]
xxviii. 1960. pp.viii.182+viii.177+viii.183+viii.185. [15,000.]

xxix. 1961. pp.viii.183+viii.201+viii.186+ viii.198. [15,000.]

xxx. 1962. pp.vii.204+ +vii.195+ vii.196. [15,000.]

xxxi. 1963. pp.vii.187+

*in progress; in vol.xiii the date is that of publication*

CATALOGUE of the Central library for animal diseases (text-books published since 1900). Royal college of veterinary surgeons: 1936. pp.46. [750.]

— Supplement. [1937]. pp.[2]. [45.]

CATALOGUE de la bibliothèque. École nationale vétérinaire: Lyon 1937. pp.167. [4000.]

B. NOYER, Bibliographie analytique des travaux scientifiques en Indochine, 1939–41. Sciences médicales et vétérinaires. Hanoï 1943. pp.48. [569.]

A MAGYARORSZÁGRA érkezett szovjet mezőgazdasági irodalom katalógusa. Mezőgazdasági tudományok tájékoztató intézete [&c.]: [Budapest] 1948. pp.48. [500.]

[RAMÓN PONCE PAZ], Contribución a la bibliografía veterinaria americana. Universidad de San Marcos: Facultad de medicina veterinaria: Lima 1951. ff.47. [1100.]*

FRANCO VERGANI and CARLOS DÍAZ UNGRÍA, Bibliografía veterinaria venezolana. Instituto de investigaciones veterinarias (vol.v, no.21): Caracas 1953. pp.[ii].715–1899. [7006.]

ALEKSANDER PERENC, Pierwsze druki weterynaryjne w Polsce. Universitas Mariae Curie Skłodowska: Annales: Sectio DD: Supplementa (no.1): Lublin 1955. pp.115. [15.]

CATALOGUE of modern works 1900–1954, with a section showing periodicals and reports. [Second edition]. Royal college of veterinary surgeons: Memorial library: 1955. pp.[v].98. [1000.]
— Supplement 1. [1957]. pp.16. [200.]
— Supplement 2. [1959]. pp.14. [250.]

RECHERCHES scientifiques de médecine vétérinaire publiées de 1904 à 1957. Institut Pasteur d'Algérie: Alger [*c.*1957]. pp.14. [150.]

RÉPERTOIRE des laboratoires de recherches médicales, pharmaceutiques, vétérinaires. Office national des universités et écoles françaises: 1958. pp. [ii].528. [2500.]

DOROTHEE RÜHLMANN, Bibliographie des deutschsprachigen schrifttums zur erforschung und bekämpfung von anzeigepflichtigen tierseuchen. Eine zusammenstellung der von 1935–

1956 erschienenen monographien, dissertationen und zeitschriftenaufsätze. Schriften zum bibliotheks- und büchereiwesen in Sachsen-Anhalt (no.17): Leipzig 1960. pp.xx.482. [7111.]

**Vibrio fetus.**

S[UMMER] M[ARTIN] MORRISON, Vibrio fetus. A bibliography. Colorado state university: Experiment station: Technical bulletin (no.68): Fort Collins 1960. pp.39. [571.]*

**Wheat.**

BRIEF list of references on wheat rust (excepting U.S. government publications). Library of Congress: [Washington] 1917. pp.2. [21.]*

WHEAT breeding bibliography. Imperial bureau of plant genetics: Cambridge [1931]. ff.[ii].30+[ii].31–61. [400.]*

LE BLÉ et les autres céréales (1929–1930). Université de Bordeaux: Institut d'études et de documentation économiques et sociales: Bulletin bibliographique (ser. A, no.13): Bordeaux 1931. ff.13. [300.]

LOUISE O[LDHAM] BERCAW, Wheat: cost of production, 1923–1930. References relating to the

United States and some foreign countries. Department of agriculture: Bureau of agricultural economics: Agricultural economics bibliography (no.33): Washington 1931. pp.[v].33. [110.]*

MECHANICAL peeling of wheat. Science library: Bibliographical series (no.54): [1932]. ff.[i].3. [44.]*

BIBLIOGRAPHY of baking quality tests (with particular reference to tests for small samples for use by wheat breeders). Imperial bureau of plant genetics: School of agriculture: Cambridge 1934. pp.68. [750.]

LE BLÉ (de 1933 à 1937). Université de Bordeaux: Institut d'études et de documentation économiques et sociales: Bulletin bibliographique (no. 9): Bordeaux 1937. ff.17. [450.]

**Wine.**

1. *Bibliographies*

JOSEF JULIUS MERZ, Die bibliographie und literaturgeschichte des weines. Wien [1947]. pp.8. [50.]

2. *Periodicals*

PERIODICALS on viticulture and the manufacture

of wines. Science library: Bibliographical series (no.181): 1935. ff.[i].2. [52.]*

3. *General*

CHR. W. J. GATTERER, Literatur des weinbaues aller nationen. Heidelberg 1832. pp.vi.74. [250.]

E. WAGENMANN, Catalog der oenologischen bibliothek von dr. Ad. Blankenhorn. Mittheilungen des Blankenhorn'schen oenologischen instituts Carlsruhe-Blankenhornsberg (vol.xlii): Heidelberg [*c*.1860]. pp.33. [325.]

— — Zweiter nachtrag zur oenologischen literatur-zusammenstellung. Revidirt von A. Blankenhorn. [*c*.1870]. pp.26. [475.]

[E. WAGENMANN], Bibliotheca oenologica. Heidelberg 1875. pp.[ii].98. [1500.]

VERZEICHNIS sämmtlicher schriften über weinbereitung, weinverbesserung, kellereiwesen etc. welche von 1865–1881 im deutschen buchhandel erschienen sind. Gracklauer's fachkatalog (no.31): Leipzig 1881. pp.8. [75.]

ABRIDGMENTS of specifications relating to brewing, wine-making, and distilling alcoholic liquids. Commissioners of patents.

[i]. 1634–1866. 1881. pp.xv.552. [750.]

[ii. 1867–1876. *not published.*]

iii. 1877–1883. 1887. pp.xii.264. [500.]

*no more published.*

GIACOMO SORMANNI, Catalogo ragionato delle opere di viticoltura ed enologia pubblicate in Italia o in italiano. Milano 1883. pp.[iii].139. [1250.]

RÈGLEMENT et catalogue de la bibliothèque. Société d'horticulture, d'arboriculture et de viticulture des Deux-Sèvres: Niort [printed] 1889. pp.44. [400.]

VASILY [EGOROVICH] TAIROV, Библіографическій указатель книгъ, брошюръ и журнальныхъ статей по виноградству и виноделію напечатанныхъ съ 1755 по 1890 годъ включительно. С.-Петербургъ 1891. pp.[iii]. viii.196. [4265.]

ANDRÉ L[OUIS] SIMON, Bibliotheca vinaria. A bibliography of books and pamphlets dealing with viticulture, wine-making, distillation, the management, sale, taxation, use and abuse of wine and spirits. 1913. pp.viii.340. [6500.]

*180 copies printed.*

LIST of references on wine and the wine industry. Library of Congress: [Washington] 1920. ff.9. [104.]*

VAJEN E[ILLEEN] HITZ, The grape industry. A selected list of references on the economic aspects of the industry in the United States, 1920–1931. Department of agriculture: Bureau of agricultural economics: Agricultural economic bibliography (no.36): Washington 1932. pp.[ii].ii.161. [416.]*

WINE and other beverages. Compiled by the ... Wine trade club. Third edition. National book council: Book list (no.70): 1932. pp.[2]. [60.]

— Fourth edition. Wine and food. Compiled by André L. Simon. 1941. pp.[4]. [125.]

BOOKS on viticulture and the manufacture of wines, 1880–1934. Science library: Bibliographical series (no.182): 1935. ff.[i].8. [182.]*

— Supplement. Books on viticulture, 1935–1953.... (no.722): 1953. ff.6. [122.]*

SELECTIVE bibliography of wine books. Wine institute: [San Francisco 1944]. ff.40. [400.]*

MAYNARD A[NDREW] AMERINE and LOUISE B. WHEELER, A check list of books and pamphlets on grapes and wine and related subjects, 1938–1948. Berkeley &c. 1951. pp.v.240. [1789.]

PATRICIA DICEY, Wine in south Africa. A select bibliography. University of Cape Town: School of librarianship: Bibliographical series: [Capetown] 1951. ff.29. [159.]*

GUY PRINCE, Current books about wine. National book league: 1955. pp.[ii].14. [50.]

JÓZSEF BAKOS, Tokajhegyalja és bortermelése néprajzi, hely-, gazdaságés művelődéstörténeti irodalma. A Sárospataki Rákóczi muzeum füzetei (no.4–7):Sárospatak 1957. pp.[iii].coll.172. [1750.]

N. G. ARAMYAN, Библиографический указатель отечественной литературы по технологии вина, 1948–1956. Главное управление сельскохозяйственной науки [&c.]: Ереван 1957. pp.128.

MAYNARD A[NDREW] AMERINE, A short check list of books and pamphlets in english on grapes, wines and related subjects, 1949–1959. [San Francisco 1959]. pp.[iii].61. [600.]*